Statistical Process Control Methods for Long and Short Runs

Also available from ASQC Quality Press

Glossary and Tables for Statistical Quality Control
ASQC Statistics Division

SPC Tools for Everyone
John T. Burr

Pyzdek's Guide to SPC, Volume 1: Fundamentals
Thomas Pyzdek

Pyzdek's Guide to SPC, Volume 2: Applications and Special Topics
Thomas Pyzdek

Introduction to Statistical Quality Control, Second Edition
Douglas C. Montgomery

SPC for Practitioners: Special Cases and Continuous Processes
Gary Fellers

Defect Prevention: Use of Simple Statistical Tools
Victor E. Kane

To request a complimentary catalog of publications, call 800-248-1946.

Statistical Process Control Methods for Long and Short Runs

Second Edition

Gary K. Griffith

ASQC Quality Press
Milwaukee, Wisconsin

Statistical Process Control Methods for Long and Short Runs, Second Edition
Gary K. Griffith

Library of Congress Cataloging-in-Publication Data
Griffith, Gary.
Statistical process control methods for long and short runs / Gary K. Griffith. — 2nd ed.
p. cm.
Includes bibliographical references and index.
ISBN 0-87389-345-X
1. Process control—Statistical methods. I. Title. II. Title: Statistical process control methods.
TS156.8.G75 1996
670.42'7—dc20 95-41877
CIP

10 9 8 7 6 5 4 3 2 1

ISBN 0-87389-345-X

Acquisitions Editor: Susan Westergard
Project Editor: Jeanne W. Bohn

ASQC Mission: To facilitate continuous improvement and increase customer satisfaction by identifying, communicating, and promoting the use of quality principles, concepts, and technologies; and thereby be recognized throughout the world as the leading authority on, and champion for, quality.

Attention: Schools and Corporations
ASQC Quality Press books, audio, video, and software are available at quantity discounts with bulk purchases for business, educational, or instructional use. For information, please contact ASQC Quality Press at 800-248-1946, or write to ASQC Quality Press, P.O. Box 3005, Milwaukee, WI 53201-3005.

For a free copy of the ASQC Quality Press Publications Catalog, including ASQC membership information, call 800-248-1946.

Printed in the United States of America

 Printed on acid-free paper

ASQC
Quality Press
611 East Wisconsin Avenue
Milwaukee, Wisconsin 53202

To my family—Sharon, Christopher, Gary, Brian, and Kimberlee—
for their patience, understanding, and support in this project.

Contents

Preface

The purpose of this book is to provide the reader with a variety of statistical process control (SPC) tools for various industrial applications and to enhance the learning process through direct steps, practice problems, and solutions. With this purpose in mind, this book does not cover the in-depth supporting statistics for each of these methods, but focuses on how to use them for process control. There are many other statistical texts available that can be used for understanding the statistical basis behind these methods. This book can be used by machinists, inspectors, engineers, quality managers, buyers, SPC facilitators, and others who want to enhance their knowledge of SPC methods.

The improvements that have been made in this second edition include the following:

- New coverage on process control planning methods
- Expanded coverage on short-run variables and short-run attributes charts
- Expanded coverage on gage repeatability and reproducibility studies
- Additional appendices (including *F* tests)
- Expanded coverage on regression analysis (scatter diagrams)
- Additional coverage on problem solving and problem-solving tools
- Clearer artwork
- Additional chapter review questions

I sincerely hope that the use of this book will contribute to improved quality and productivity.

Acknowledgments

There are numerous examples of control charts in this book where the practice charts have been prepared manually, followed by an example of the same chart shown using SPC software. I am grateful to CIM Vision International of Torrance, California, for the permission to use these examples from its Statistical Quality Manager (SQM) software product.

Comparing these software examples to the manually created charts, you will notice that values for centerlines, control limits, capability indexes, and other statistical results may be slightly different than the manually computed values. The primary reason for this mathematical difference is that the manual charts were prepared with data rounded to one more decimal place, and the software-prepared charts are in double precision (for example, 14-decimal place accuracy). Also, note that all manually prepared charts were prepared after the subgroups were all taken, and some of the limits of the software-prepared charts were computed in real time. The medians control chart example is also different because I chose to produce it in the traditional manner (all data points plotted) assuming that a range chart may not be used. The SQM software example plots only the medians because there is always an optional range chart to monitor variation.

CHAPTER 1

Introduction to Statistical Process Control

INTRODUCTION

Statistical process control (SPC) is a tool that applies basic statistics to control processes. A *process* is any collection of people, material, methods, equipment, measurement, and environment that produces a certain output.

All processes are subject to variation, and no two processes are alike. Even though the difference between the sizes of two machined parts may be small, they are different. This variation is the reason specifications have tolerances.

Even though common causes of variation are difficult to identify and correct, the output of the process is predictable. Some examples of common causes of variation are vibration, heat, and humidity.

A process is not predictable if special causes of variation exist. Special causes of variation can usually be found and corrected, particularly if you know when they are occurring. Some examples of special causes of variation include a broken tool, tool wear, a power surge, or loose fixtures. At times, even special causes of variation are difficult to find.

Some refer to the total variation of a process as the *natural* tolerance of the process. Keep in mind that a process will produce parts according to its own inherent variation regardless of the product specification. The important thing is to control the process, then make sure that the natural tolerance (process variation) is less than the specification tolerance, which defines whether or not the product is acceptable. It is intended that most parts produced will be near nominal size. This is called process capability and will be discussed in chapter 7.

The primary goal of SPC is to get the process in control (or repeating its inherent variation) by eliminating all of the special causes of variation. A process that is in control will produce parts consistently (or be more predictable) because special causes of variation have been eliminated. Once a process is in statistical control, it can then be compared to the specification limits to see if it is capable of producing quality products. The capability of a process is directly related to the ability of the process to produce parts within specification limits (refer to chapter 7).

RUN CHARTS

A *run chart* is a graph of measurements from a process over time. Using run charts, the operator can begin to understand process variation and identify some special causes of variation (such as tool wear). At times, run charts have been started as a preliminary control method until sufficient data have been collected to convert them to statistical control charts (with statistical control limits). Consider the data in Table 1.1.

With no visibility of the pattern of variation over time, an operator could make adjustment errors by only looking at raw data values as they are measured. Graphing these values (as shown in Figure 1.1) provides a better understanding of the variation of the process. Run charts are limited in identifying special causes because they do not have control limits. In any case, a run chart of the data provides better visibility on the process and can help avoid adjustment errors such as overadjustment or underadjustment.

Overadjustment is changing a process when it does not need it. *Underadjustment* is not changing process when it does need to be changed. The run chart in Figure 1.1 shows an upward trend in the data from the process, so the process would have to be adjusted. Without the chart, the operator may adjust the process on the third part produced since it is much closer to the upper specification limit than the first part. In this case, the operator using the run chart would wait until a trend appears before making the necessary adjustment.

DEFINITION OF A PROCESS IN CONTROL

A process is in statistical control when all special (identifiable) causes of variation have been eliminated and only common (chance) causes exist. This is evidenced by plot points on control charts remaining within statistical control limits and exhibiting random variation between those limits (Figure 1.2). Note, however, that a process in control may not necessarily be capable of meeting specifications. The common cause system may have an overall natural variation that exceeds specification boundaries, therefore, the process is in control but not capable. For further information on this, refer to chapter 6.

DEFINITION OF A CAPABLE PROCESS

A process is considered to be capable when its inherent variation is within part specification boundaries by a predetermined margin (often measured by capability indexes discussed in chapter 7). The most popular capability indexes today, such as the C_p and C_{pk} (covered in

Table 1.1. Sample data.

Tolerance limits	Measurements			
Diameter .192–.202	(1)	.195	(9)	.196
	(2)	.196	(10)	.199
	(3)	.199	(11)	.197
	(4)	.198	(12)	.197
	(5)	.194	(13)	.200
	(6)	.196	(14)	.198
	(7)	.195	(15)	.197
	(8)	.197	(16)	.198

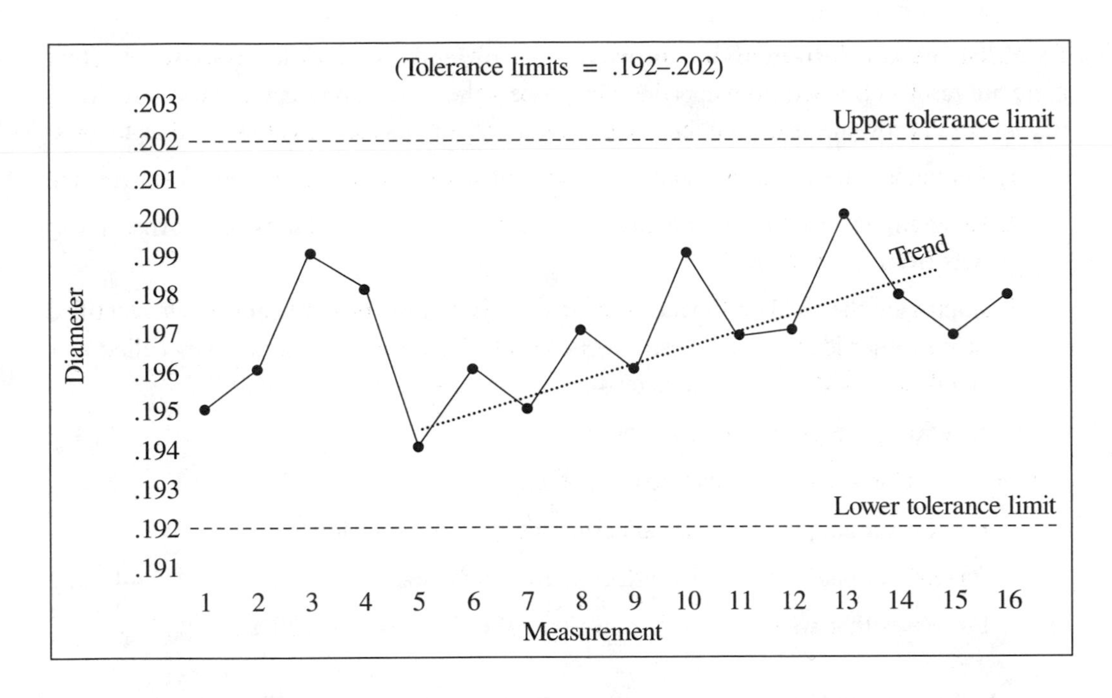

Figure 1.1. A run chart for the data.

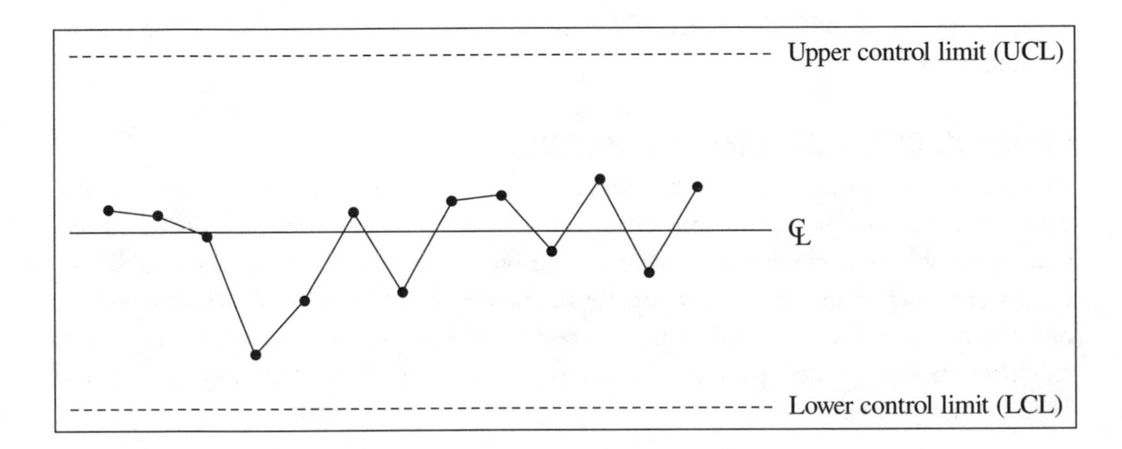

Figure 1.2. A sample process control chart.

chapter 7), require a specific ratio of the total part specification tolerance versus the total variation of the process. The natural tolerance of the process is represented by the plus and minus three sigma limits from the mean (or the six-sigma spread). Chapter 7 explains process capability studies and indexes that are most widely used in industry today.

PLANNING FOR SPC

The following elements of planning for SPC apply in traditional or short-run SPC methods. A review of these elements will help make the implementation of SPC more effective.

Establish quality/productivity improvement objectives. In order to make sure that there are results obtained from applying SPC, assess the system, product, or process to identify where improvements need to be made. Quality objectives can be categorized as follows:

1. Reducing internal failure quality costs (such as the cost of scrap, rework, or repair).
2. Reducing external failure quality costs (such as warranty claims or returned products from the customer).
3. Improving the yield on functional characteristics of the product (either self-identified or customer-identified). These functional characteristics are sometimes called key, significant, or dominant characteristics.
4. Solution of recurring quality problems.

Productivity-related objectives can be categorized as follows:

1. Process output problems that affect throughput in the factory
2. Process output problems that affect delivery schedules
3. Processes that are subject to just-in-time (JIT) inventory objectives

Correlate (or translate) objectives to products, processes, and characteristics. No matter what objective has been established for the SPC implementation, the solution lies in the product, process, or characteristic that is the root cause of the problem. It is important to do research to identify the source of the problem. SPC can then be used to help remove the source (or root cause).

PROCESS CONTROL CHARACTERISTICS

Control characteristics are specific characteristics of a system, assembly, subassembly, component part, or process parameter(s) that have a direct affect on the objective, function, or reliability of the product. There are various levels of control characteristics from the system down to the component part (or the process). All levels need to be assessed in order to be successful. The following is a review of some specific definitions of control characteristics (or process parameters) that may be helpful when completing this element of SPC implementation.

- *Key product characteristic* is a characteristic of the product that affects fit, form, function, or reliability.

- *Customer exciter* is a characteristic of the product or service that stands out from all other characteristics as one that pleases the customer (when it is correct) or upsets the customer (when it is defective). Customer exciters are not always fit, form, or functional characteristics. Sometimes, the only way to identify a customer exciter is to ask the customer. In other cases, customer complaint files can be reviewed to identify these characteristics. In any case, this approach should be taken because it can mean the difference between getting or losing the job.

- *Response variables* are variables of the output of a process that respond or are affected by problems within the process. Technically, all characteristics of the output of a process are response variables, but this section is intended to suggest that people should focus on single variables of a process that respond to many different problems in the process.

A machine shop example is surface finish on a machined part. Surface finish is a characteristic of the product that is affected by process parameters or problems such as incorrect speed and feed rates, cutting tool wear, wrong or no coolant, and other process factors. Therefore, if surface finish is plotted on a control chart, it will identify when any one of these factors is a problem. A nice side effect is the fact that dimensional control is also affected by some of the same factors that cause surface finish problems. Hence, improvements in surface finish will also provide degrees of improvement in dimensional control.

- *Key path variables* is the author's term for two types of variables that are worthy for consideration as key variables for charting. *Key path process variables* are any variables produced at an earlier process that have a direct affect on the success of a subsequent process. For example, a diameter produced at operation 10 is used in operation 60 as a locating diameter for a concentricity requirement of the finished part.

Key path machining variables are best applied in computer numerically controlled (CNC) machines where several dimensions are cut at the same time with the same cutting tool. The idea is to identify the single dimension cut by that tool where the worst variability is expected. For example, a CNC lathe cuts several diameters. In some cases, the worst variability can be seen at the largest diameter that is farthest away from the locating jaws of the lathe. If this diameter is charted and special causes of variation are eliminated, the result is variability improvement in all diameters in the path.

- *Control factors (parameters)* are specific parameters of a process that have a direct affect on the output. An example is the effect of temperature of a heat treat oven on the hardness of the output product.

IDENTIFYING CONTROL CHARACTERISTICS/PARAMETERS

The following approaches are commonly used to identify control product characteristics (or process parameters).

- *Brainstorming* is a powerful tool for a natural work team to identify control characteristics or parameters. Refer to chapter 9 for more information on brainstorming.

- *Cause-and-effect analysis* can also be used in a structured approach. Cause-and-effect analysis and brainstorming together are very effective. Refer to chapter 9 for more information on cause-and-effect analysis.

- *Failure modes and effects analysis (FMEA) on the product* can help identify product key characteristics. The method for product FMEAs is to first identify the various failure modes (ways the product can fail), then identify the effects of each of those failure modes on the end product. This is often done from a top-down point of view.

- *FMEA on the process* can help identify control process variables (parameters). The method for process FMEAs is to first identify the various failure modes (ways the process can fail to perform), then identify the effects of each of those failure modes on the output of the process. Refer to other books for more detailed information on FMEA.

- *Statistical design of experiments* is a very powerful tool for identifying control factors of a process, especially when multiple variables are involved. Design of experiments is most helpful toward understanding the factors of a multifactor process. The following general objectives apply to designed experiments.

a. Identify the input factors that have the main affect on the output response for control purposes.

b. Identify the levels (or settings) of input factors that help to make the response variable robust (or not affected by) uncontrollable factors.

c. Identify the level of input factors that reduce the variation in the output response variable.

d. Identify the levels (or settings) of input variables that produce a nominal out response.

Refer to appropriate textbooks for more detailed information on statistical design of experiments.

THE PROCESS CONTROL PLAN

Once the control characteristics or parameters have been identified, a process control plan (as shown in Figure 1.3) should be prepared. The control plan is the key for effective process control and SPC training. The control plan is the form that completes SPC planning and communicates SPC technical requirements to applicable personnel. The plan, of course, should be reviewed to ensure to see that it meets the planned objectives for SPC and is technically accurate. Figure 1.3 is an example of a short-run process control plan that focuses on a lathe that produces input shafts. The key elements of any control plan (short-run or traditional) are covered in this example.

A process control plan communicates all of the information required to perform SPC on the process. This information includes the process to be controlled, the control characteristics, the type of chart to use, sample sizes and frequencies, the measurement or gaging device, and other pertinent information.

Notice in Figure 1.3 that this short-run SPC control plan applies to aluminum and steel input shafts and that they must be charted separately. The basis for separate charting is that the aluminum shafts are expected to vary differently than the steel shafts. This short-run control plan, when implemented properly, will provide statistical control on the lathe process and will make the lathe capable of producing *all* part numbers of input shafts within their specifications. Every action taken on out-of-control causes will have a positive affect on all future part numbers of input shafts.

HOME RUN SPC

Home run SPC (see Figure 1.4) can be described as follows:

First base: The action taken on the process improves the yield of the process up to or beyond the applicable capability (for example, C_{pk}) goal.

Second base: The same improvement action taken in the process also reduced manufacturing costs.

Third base: The same improvement action taken increased production rate (or reduced cycle time). Hence, improved throughput.

Home run: The same improvement action taken on process also took a future process, product, assembly, or system to one or more bases.

DoRight Mfg. Co.

Process: Lathe	**Part name:** Input shafts		**Part no.:** All
Next assembly: Pilot housings	**Prepared by:** J. Smith	**Approved by:** H. James	**Date prepared:** 2-2-95
Team leader: K. Samuels	**Title:** SPC facilitator	**Title:** Mfg. manager	**Date revised:** N/A

Control characteristics

A.	Outside diameter	**D.**		**G.**	
B.	Runout to datum -A-	**E.**		**H.**	
C.		**F.**		**I.**	

Process control plan

Control characteristics					Measurements		Process variation		
Char.	**Specification limits**	**Statistical method**	**Sample size and frequency**	**Initial C_{pk}**	**Gage/ method**	**Gage R&R**	**Process step/ operator number**	**Key process parameter and setting**	**Control method**
A	Various	Target avg. and range ($N = 0$) **(1)**	2/Hour	Worst 1.2	Micrometer (.0001″)	8%	30	Speed/feed	CNC
B	Various	Target avg. and range ($H = 0$) **(1)**	2/Hour	Worst 1.1	V-Block and indicator (.0001″)	11%			

(1) Two different charts (one for aluminum shafts and one for steel shafts)

Figure 1.3. An SPC control plan.

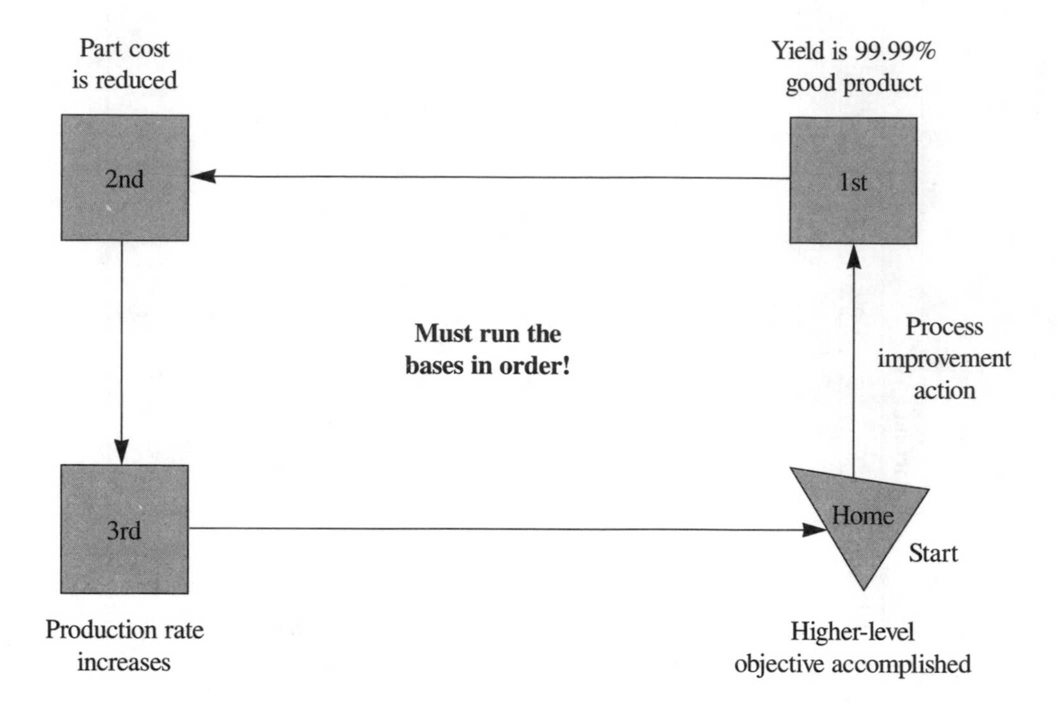

Figure 1.4. An example of home run SPC.

The idea of home run SPC puts quality (yield) first, but does not limit SPC objectives to only quality improvement. The one rule to home run SPC is, as in baseball, the bases must be run in order. First base is quality because improvements in quality automatically provide a certain level of cost, schedule, inventory, and/or profit improvements. The difference between home run SPC and baseball is that, in baseball, you can only run the bases once. Using home run SPC, the bases can be run several times on one hit. SPC can be a quality *and* productivity improvement tool. When applying short-run techniques, duplicate home runs can be achieved because many part numbers are affected by a single process improvement action.

BASIC STATISTICS REVIEW

The following is an introduction to the symbols and basic statistics that are used in this book and examples of each.

X_i—Represents an individual measurement or data point.

Example .501 or .750

$\bar{X}$ (pronounced X-bar)—Represents the average of several measurement or data points.

Example 5 6 7 8 9

$$\bar{X} = \frac{\Sigma X_i}{n} = \frac{5 + 6 + 7 + 8 + 9}{5} = 7$$

where

Σ = sum

n = sample size

X_i = individual measurements

$\bar{X}_m$—Represents the moving average of several individual measurements. The moving average is the average of every n set of measurements, dropping the first measurement each time.

Example 5 6 7 8 9

$\bar{X}_m$, where $n \neq 3$, is the average of 5, 6, and 7; and then 6, 7, and 8; then 7, 8, and 9; and so on.

R—Represents the range of the measurements. The range is the largest measurement minus the smallest measurement.

Example 5 6 7 8 9

where

R = The largest data point minus the smallest

$R = 9 - 5$

$R = 4$

$\bar{R}$ (pronounced R-bar)—Represents the average range. It is the average of several ranges.

Example $R = 5$ $R = 6$ $R = 7$ $R = 8$ $R = 9$

$$\bar{R} = \frac{\Sigma R}{k} = \frac{5 + 6 + 7 + 8 + 9}{5} = 7$$

where k = the number of ranges

R_m—Represents the moving range. The moving range is the range of n measurements dropping the first measurement each time.

Example 5 6 7 8 9 ($n = 2$)

First calculate the range of 5 and 6, then 6 and 7, then 7 and 8, and so on.

s—Represents the sample standard deviation of data. Refer to the section on average and standard deviation charts (page 19) for examples. Note: Some calculators use the button σ_{n-1} to represent the sample standard deviation.

σ (sigma)—Represents the population standard deviation. Note: Some calculators use the button σ_n to represent the population standard deviation.

$\bar{s}$ (sigma bar)—Represents the average standard deviation. It is the average of several sample standard deviations.

Example $s_1 = 2$ $s_2 = 3$ $s_3 = 4$ $s_4 = 6$

$$\bar{s} = \frac{\Sigma s}{k} = \frac{2 + 3 + 4 + 6}{4} = \frac{15}{4} = 3.75$$

where k = the number of standard deviations

$\hat{\sigma}$ (called sigma hat)—Represents the estimated standard deviation and is used with capability studies associated with range (or standard deviation) control charts.

Example 1 $\bar{R} = 6 \quad n = 5$

$$\hat{\sigma} = \frac{\bar{R}}{d_2} = \frac{6}{2.326} = 2.58$$

Example 2 $\bar{s} = 6 \quad n = 10$

$$\hat{\sigma} = \frac{\bar{s}}{c_4} = \frac{6}{.973} = 6.17$$

where d_2 and c_4 are constant factors (refer to the table of factors for control charts in Appendix B). These factors depend on the sample size (n).

M_D—Represents the median or midpoint of ordered data (or measurements).

Example Data = 5 8 4 7 3

Ordered = 3 4 5 7 8

M_D = 5 (the value in the middle)

p—Represents the fraction nonconforming of the process. This value is one plotted point on a p chart.

Example 50 parts inspected

5 parts are nonconforming

$$p = \frac{\text{The quantity of nonconforming}}{\text{The quantity inspected}} = \frac{5}{50} = .10$$

$\bar{p}$ (P-bar)—Represents the average fraction nonconforming. This value is the centerline of a p chart.

Example Quantity nonconforming 5 6 12 3

Quantity inspected 50 50 50 50

$$\bar{p} = \frac{\text{Sum of all nonconforming}}{\text{Sum of all inspected}} = \frac{5 + 6 + 12 + 3}{200} = .13$$

np—Represents the number of nonconformances (defectives) found in a subgroup.

$n\bar{p}$ (np-bar)—Represents the average number of nonconformances found.

Example Nonconformances 20 19 8

$$n\bar{p} = \frac{\text{Sum of nonconformances}}{k} = \frac{47}{3} = 15.7$$

$100p$—Represents the percent nonconforming

Example

$$100p = \frac{\text{Quantity nonconforming}}{\text{Quantity inspected}} \times 100$$

$$100p = \frac{5 \text{ nonconformances}}{50 \text{ inspected}} \times 100 = 10\%$$

$100\bar{p}$—Represents the average percent nonconforming (defective).

$$100\bar{p} = \frac{\text{Sum of all nonconformances}}{\text{Sum of all inspected}} \times 100$$

Example

Nonconformances	5	6	12	3
Inspected	50	50	50	50

$$100\bar{p} = \frac{5 + 6 + 12 + 3}{200} \times 100 = 13\%$$

c—Represents the number of nonconformities (defects) found.

Example 15 nonconformities (defects) found on one part

$\bar{c}$ (C-bar)—Represents the average number of nonconformities found.

Example Nonconformities found: 5 12 6 9

$$\bar{c} = \frac{\text{Sum of nonconformities}}{k} = \frac{5 + 12 + 6 + 9}{4} = \frac{32}{4} = 8$$

u—Represents the number of nonconformities per unit.

$$u = \frac{\text{Number of nonconformities found}}{\text{Number of units inspected}}$$

Example 17 units inspected

28 nonconformities found

$$u = \frac{28}{17} = 1.6$$

$\bar{u}$ (U-bar)—Represents the average number of nonconformities per unit.

$$\bar{u} = \frac{\text{Sum of all nonconformities found}}{\text{Sum of all units inspected}}$$

Example

Nonconformities found:	9	20	8	3
Units inspected:	20	20	20	20

$$\bar{u} = \frac{9 + 20 + 8 + 3}{80} = \frac{40}{80} = 0.5$$

CHAPTER 2

Traditional Variables Control Charts

INTRODUCTION

The best tools for process control are statistical control charts. Statistical charts are used for

1. Detecting special causes of variation in the process at the time they exist.
2. Measuring the natural tolerance of the process, which is the result of common causes of variation.
3. Assistance in getting the process in control and capable of meeting specifications consistently.

Variables are measured characteristics of the part such as weight, length, diameters, and torque. Variable charts are more sensitive to variation in the process than attribute charts and should be considered at all times when measurements can be made on the product.

AVERAGE AND RANGE CONTROL CHARTS

The purpose of average and range control charts is to control the level and variability of a process. Average and range charts are traditionally used in mass production operations. In general, an average and range chart requires a minimum of 20 subgroups just to start. The formulas for average and range charts follow.

Range (R) charts

Centerline is $\bar{R}$

$$\bar{R} = \frac{\Sigma R}{k} = \frac{\text{Sum of all ranges}}{\text{Total number of ranges}}$$

Upper control limit (UCL) $= D_4 \times \bar{R}$

Lower control limit (LCL) $= D_3 \times \bar{R}$

Average ($\bar{X}$) charts

Centerline is $\bar{\bar{X}}$

$$\bar{\bar{X}} = \frac{\Sigma \bar{X}}{k} = \frac{\text{Sum of } \bar{x} \text{ values}}{\text{Number of } \bar{x} \text{ values}}$$

$$\text{UCL} = \bar{\bar{X}} + [A_2 \bullet R]$$
$$\text{LCL} = \bar{\bar{X}} - [A_2 \bullet R]$$

The factors A_2, D_4, and D_3 are constant numbers that are found in a table and depend on the subgroup sample size. A subgroup consists of one sample taken from the process. The sample size is the number of items inspected in the sample. The table of factors for control charts is included in Appendix B. For example: If the sample size on a control chart is 5, then $A_2 = .577$, $D_4 = 2.114$, and $D_3 = 0$.

Steps to Construct an Average and Range Chart

Refer to Figures 2.1 and 2.2 for practice. Follow these steps to create your own $\bar{X}$–R chart.

1. *Purpose:* Decide the purpose of the chart. All charts have a purpose—from reducing defectives to monitoring the process. The primary purpose is process control and capability for characteristics that are related to an improvement objective.

2. *Characteristic:* It is wise to select a characteristic of the part that is important because of safety, fit, form, function, or a characteristic that has high scrap, rework, or repair rates. It is also wise to select a characteristic of a part that will focus on the stability of the process or will provide simultaneous control of several characteristics (home run SPC).

3. *Specifications:* Make sure that the characteristic is clearly defined in the specification so that there is no misunderstanding in measurement technique, acceptance decisions, or evaluation.

4. *Measurement:* The measurement or evaluation of the characteristic is important. Measurements will cause a control chart to control the process or cause production to overcontrol or undercontrol a process. Basic measuring rules should be followed such as the 10 percent discrimination rule and proven methods or techniques. The 10 percent discrimination rule (used the world over) is that the stated discrimination of a measuring tool should not exceed 10 percent of the total specification tolerance being measured. The measurement method or technique should also be consistent among all observers.

5. *Related information:* Fill in the part and process information at the top of the control chart.

6. *Sample sizes:* Select the sample size for the subgroups and the time interval between subgroups. At first you may take samples at close intervals and then widen them later on. Refer to chapter 11 for assistance regarding sample size, frequencies, and rational subgrouping.

7. *Data collection:* Collect a minimum of 20 subgroups of each from the process, and record them on the chart. The sample size can be any sample size, but for average and range charts, the sample size is usually between two to five samples per subgroup.

8. *Summation of subgroups:* Add the measurements in each subgroup and record the sum in the sum column.

9. *Averages and ranges:* Find the average and range for each subgroup and record them in the appropriate spaces.

10. *The grand average* ($\bar{\bar{X}}$): Calculate the grand average, which is the sum of all the subgroup averages divided by the total number of averages (or the sum of all the measurements divided by the number of measurements).

DoRight Mfg. Co.

Process: *Mill*

Variable Control Chart
Average and Range

Product: *Mount assembly*

Characteristic: *Gap dim. .5–.9*

Data	1	2	3	4	5	6	7	8	9	10	11	12	13	14	15	16	17	18	19	20	21	22	23	24	25
1	.65	.75	.75	.60	.70	.60	.75	.60	.65	.60	.80	.85	.70	.65	.90	.75	.75	.75	.65	.60	.50	.60	.80	.65	.65
2	.70	.85	.80	.70	.75	.75	.80	.70	.80	.70	.75	.75	.70	.70	.80	.80	.70	.70	.65	.60	.55	.80	.65	.60	.70
3	.65	.75	.80	.70	.65	.75	.65	.80	.85	.60	.90	.85	.75	.85	.80	.75	.85	.60	.85	.65	.65	.65	.75	.65	.70
4	.65	.85	.70	.75	.85	.85	.75	.75	.85	.80	.50	.65	.75	.75	.75	.80	.70	.70	.65	.60	.80	.65	.65	.60	.60
5	.85	.65	.75	.65	.80	.70	.70	.75	.75	.65	.80	.70	.70	.60	.85	.65	.80	.60	.70	.65	.80	.75	.65	.70	.65
Sum	3.50	3.85	3.80	3.40	3.75	3.65	3.65	3.60	3.90	3.35	3.75	3.80	3.60	3.55	4.10	3.75	3.80	3.35	3.50	3.10	3.30	3.45	3.50	3.20	3.30
Avg.																									
Range																									

Averages: .85, .80, .75, .70, .65, .60

1 2 3 4 5 6 7 8 9 10 11 12 13 14 15 16 17 18 19 20 21 22 23 24 25

Ranges: .50, .40, .30, .20, .10, 0

Figure 2.1. Practice average and range chart.

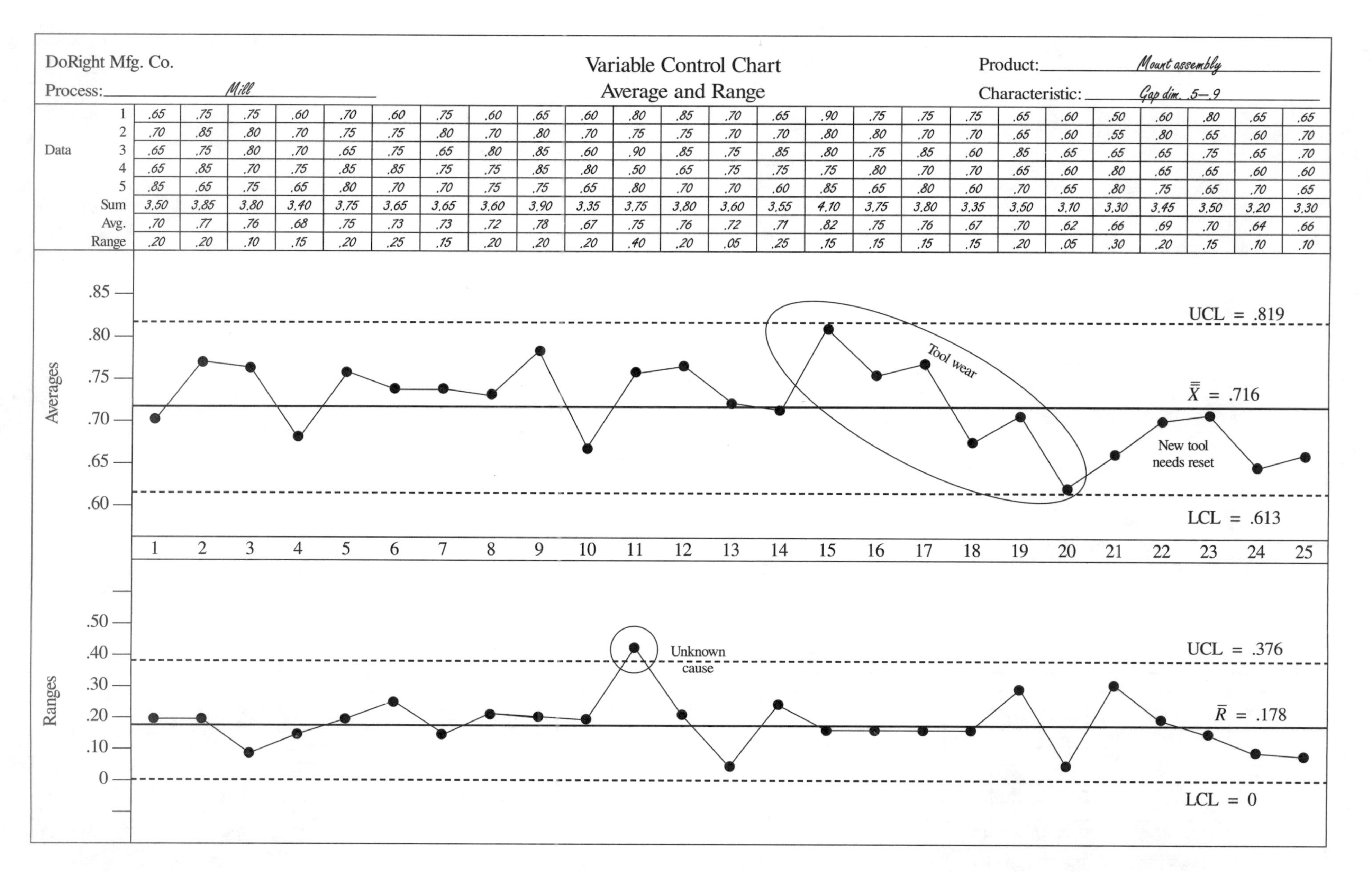

		1	2	3	4	5	6	7	8	9	10	11	12	13	14	15	16	17	18	19	20	21	22	23	24	25
Data	1	.65	.75	.75	.60	.70	.60	.75	.60	.65	.60	.80	.85	.70	.65	.90	.75	.75	.75	.65	.60	.50	.60	.80	.65	.65
	2	.70	.85	.80	.70	.75	.75	.80	.70	.80	.70	.75	.75	.70	.70	.80	.80	.70	.70	.65	.60	.55	.80	.65	.60	.70
	3	.65	.75	.80	.70	.65	.75	.65	.80	.85	.60	.90	.85	.75	.85	.80	.75	.85	.60	.85	.65	.65	.65	.75	.65	.70
	4	.65	.85	.70	.75	.85	.85	.75	.75	.85	.80	.50	.65	.75	.75	.75	.80	.70	.70	.65	.60	.80	.65	.65	.60	.60
	5	.85	.65	.75	.65	.80	.70	.70	.75	.75	.65	.80	.70	.70	.60	.85	.65	.80	.60	.70	.65	.80	.75	.65	.70	.65
	Sum	3.50	3.85	3.80	3.40	3.75	3.65	3.65	3.60	3.90	3.35	3.75	3.80	3.60	3.55	4.10	3.75	3.80	3.35	3.50	3.10	3.30	3.45	3.50	3.20	3.30
	Avg.	.70	.77	.76	.68	.75	.73	.73	.72	.78	.67	.75	.76	.72	.71	.82	.75	.76	.67	.70	.62	.66	.69	.70	.64	.66
	Range	.20	.20	.10	.15	.20	.25	.15	.20	.20	.20	.40	.20	.05	.25	.15	.15	.15	.15	.20	.05	.30	.20	.15	.10	.10

Figure 2.2. Solution to practice average and range chart.

11. *The average range* ($\bar{R}$): Calculate the average range, which is the sum of all of the subgroup ranges divided by the total number of ranges.

12. *Trial centerlines:* The grand average ($\bar{\bar{X}}$) is the centerline of the averages chart and the average range ($\bar{R}$) is the centerline of the range chart. Plot both.

13. *Trial control limits—range chart:* The range chart trial control limits are calculated as follows:

$$\text{UCL} = D_4 (\bar{R})$$

$$\text{LCL} = D_3 (\bar{R})$$

(Refer to Appendix B for factors.)

Use the UCL and the average range to select the scale for the chart. Draw the control limits in dashed lines on the range chart. If D_3 is zero, there is no LCL.

14. *Trial control limits—average chart:* The average chart trial control limits are calculated as follows:

$$\text{UCL} = \bar{\bar{X}} + [A_2 \bullet R]$$

$$\text{LCL} = \bar{\bar{X}} - [A_2 \bullet R]$$

(Refer to Appendix B for factors.)

Use the UCL and the grand average to select the scale for the chart. Draw the control limits in dashed lines on the average chart.

15. *Plotting points:* Plot each subgroup average on the average chart and each subgroup range on the range chart and connect the points.

16. *Initial decision on the chart:* Look for out-of-control conditions (see chapter 6). If there are out-of-control conditions, attempt to find the special cause. If the special cause is found, recalculate the centerline and control limits excluding the data that were out of control. If there are no out-of-control conditions, continue plotting the chart from the process.

17. *Acting on out-of-control conditions:* All out-of-control conditions on a control chart deserve some action to find the special cause and eliminate it. If the cause is found and eliminated, recalculate the control limits.

18. *Control limits:* Control limits should be reviewed with possible revision frequently (at least every 20 to 30 subgroups). As the process tends to worsen, control limits tend to widen. As the process tends to improve, control limits tend to narrow. Control limits should be revised when special causes are found and eliminated.

Remember to *always* mark the chart to indicate actions taken. Common actions include

1. Out-of-control conditions and the cause
2. Reason chart was stopped; for example, machine down
3. Reason chart was started again; for example, new setup
4. When control limits are revised
5. Any adjustments made to the process

Control charts are a running record of the process and making the chart can help production to further understand the process and its variables. See Figure 2.3 for another example of an average and range chart.

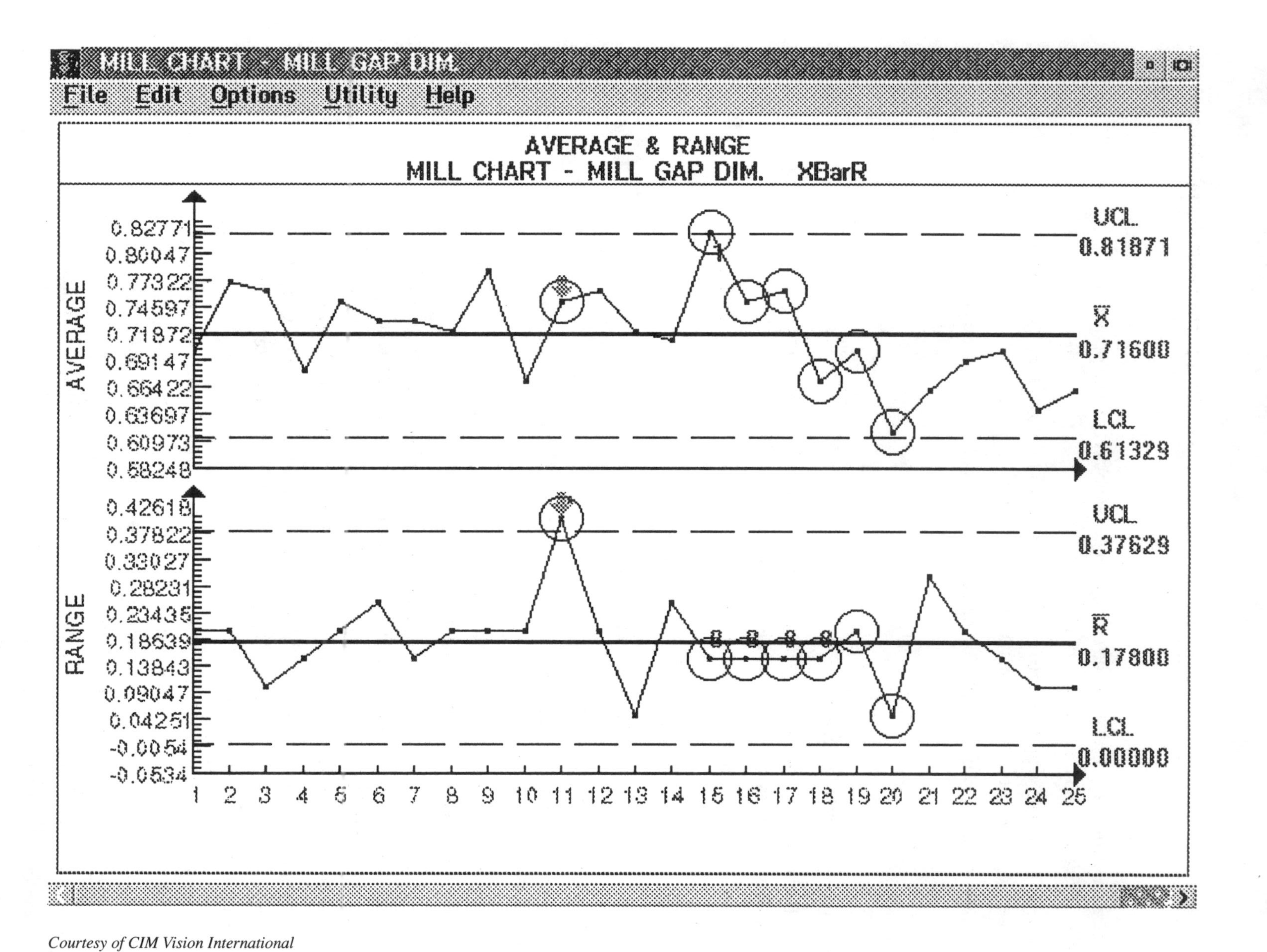

Courtesy of CIM Vision International

Figure 2.3. SQM software example of an average and range chart.

It is important to take action on processes that are not in statistical control or not capable. The table of action is Figure 2.4 shows examples of actions necessary for achieving a controlled and capable process.

AVERAGE AND STANDARD DEVIATION CHARTS

Average and standard deviation ($\bar{\bar{X}} - s$) charts are often used for increased sensitivity to variation. These charts are more sensitive than average and range charts and also are more difficult to work with because of the tedious calculation of the standard deviation when they are manually implemented. The average and standard deviation chart is used when subgroup sample sizes are 10 or more because the ranges become inefficient at samples of 10 or more. The formula for sample standard deviation(s) is

$$s = \sqrt{\frac{\Sigma (X_i - \bar{X})^2}{n - 1}}$$

where

Σ = the sum
X_i = the individual measurements
$\bar{X}$ = the average
n = the sample size
s = the sample standard deviation

Calculation of the sample standard deviation is performed in the following manner.

Example Data 5, 6, 7, 6, 8 $n = 5$ $\bar{X} = 6.4$

$$s = \sqrt{\frac{\Sigma (X_i - \bar{X})^2}{n - 1}}$$

$$= \sqrt{\frac{5.20}{4}}$$

$$= \sqrt{1.30}$$

$$s = 1.140 \text{ (rounded off)}$$

	In control	Out of control
Capable	Monitor with the control chart.	Capability should not be studied until the process is in control.
Not capable	See alternatives.	1. Get the process in control. 2. Make the process capable.

Figure 2.4. A table of action for SPC.

In most cases, a calculator/computer should be used to avoid the time-consuming mathematics. Note that the average and standard deviation charts are not as practical to use as the average and range charts when the charts are manually prepared. Because of the availability of computers and powerful calculators, however, the use of average and standard deviation charts has increased.

The charts for averages is constructed in the same way as described earlier except that when the standard deviation chart is used, the control limits are calculated using different factors. The control limits for an average chart using the standard deviation are calculated using the following formula.

$$\bar{\bar{X}} \pm [A_3 \bullet \bar{s}]$$

The standard deviation chart control limits are calculated using the following formula.

$$\text{UCL} = B_4\bar{s}$$

$$\text{LCL} = B_3\bar{s}$$

$$\bar{s} = \frac{\text{Sum of all sample standard deviations}}{\text{The total number of standard deviations}}$$

Steps to Construct an Average and Standard Deviation Chart

Refer to practice charts in Figures 2.5, 2.6, and 2.7. Note: These practice charts are using subgroup samples of five in order to save time while practicing the method. Actual average and standard deviation charts typically use sample sizes of 10 or more (as previously discussed).

1. Decide the purpose of the chart.
2. Select the characteristic to be controlled.
3. Collect and record the data.
4. Calculate the average for each subgroup.
5. Calculate the sample standard deviation for each subgroup.
6. Find the grand average ($\bar{\bar{X}}$) and the average standard deviation ($\bar{s}$).
7. Draw the centerlines on each chart.
8. Calculate the control limits for each chart.
9. Plot the averages and the standard deviations.
10. Interpret for control and revise limits as needed.
11. Continue the chart for process control.

INDIVIDUALS AND MOVING RANGE CHARTS

Individuals and moving range charts ($X_i - R_m$) use individual readings instead of subgroup averages. Some examples of applications for the $X_i - R_m$ charts are: short production runs where data are scarce, destructive testing, special process tests, and any process where individual measurements are necessary or expensive.

There are some drawbacks to using individuals and moving range charts. Use caution in interpretation if the data are not normally distributed. Also, individuals and moving

DoRight Mfg. Co.

Process: *Mill*

Variable Control Chart
Average and Standard Deviation

Product: *Mount assembly*

Characteristic: *Gap dim. .5–9*

Data	1	2	3	4	5	6	7	8	9	10	11	12	13	14	15	16	17	18	19	20	21	22	23	24	25
1	.65	.75	.75	.60	.70	.60	.75	.60	.65	.60	.80	.85	.70	.65	.90	.75	.75	.75	.65	.60	.50	.60	.80	.65	.65
2	.70	.85	.80	.70	.75	.75	.80	.70	.80	.70	.75	.75	.70	.70	.80	.80	.70	.70	.65	.60	.55	.80	.65	.60	.70
3	.65	.75	.80	.70	.65	.75	.65	.80	.85	.60	.90	.85	.75	.85	.80	.75	.85	.60	.85	.65	.65	.65	.75	.65	.70
4	.65	.85	.70	.75	.85	.85	.75	.75	.85	.80	.50	.65	.75	.75	.75	.80	.70	.70	.65	.60	.80	.65	.65	.60	.60
5	.85	.65	.75	.65	.80	.70	.70	.75	.75	.65	.80	.70	.70	.60	.85	.65	.80	.60	.70	.65	.80	.75	.65	.70	.65
Sum	3.50	3.85	3.80	3.40	3.75	3.65	3.65	3.60	3.90	3.35	3.75	3.80	3.60	3.55	4.10	3.75	3.80	3.35	3.50	3.10	3.30	3.45	3.50	3.20	3.30
Avg.																									
Sigma																									

Figure 2.5. Practice average and standard deviation chart.

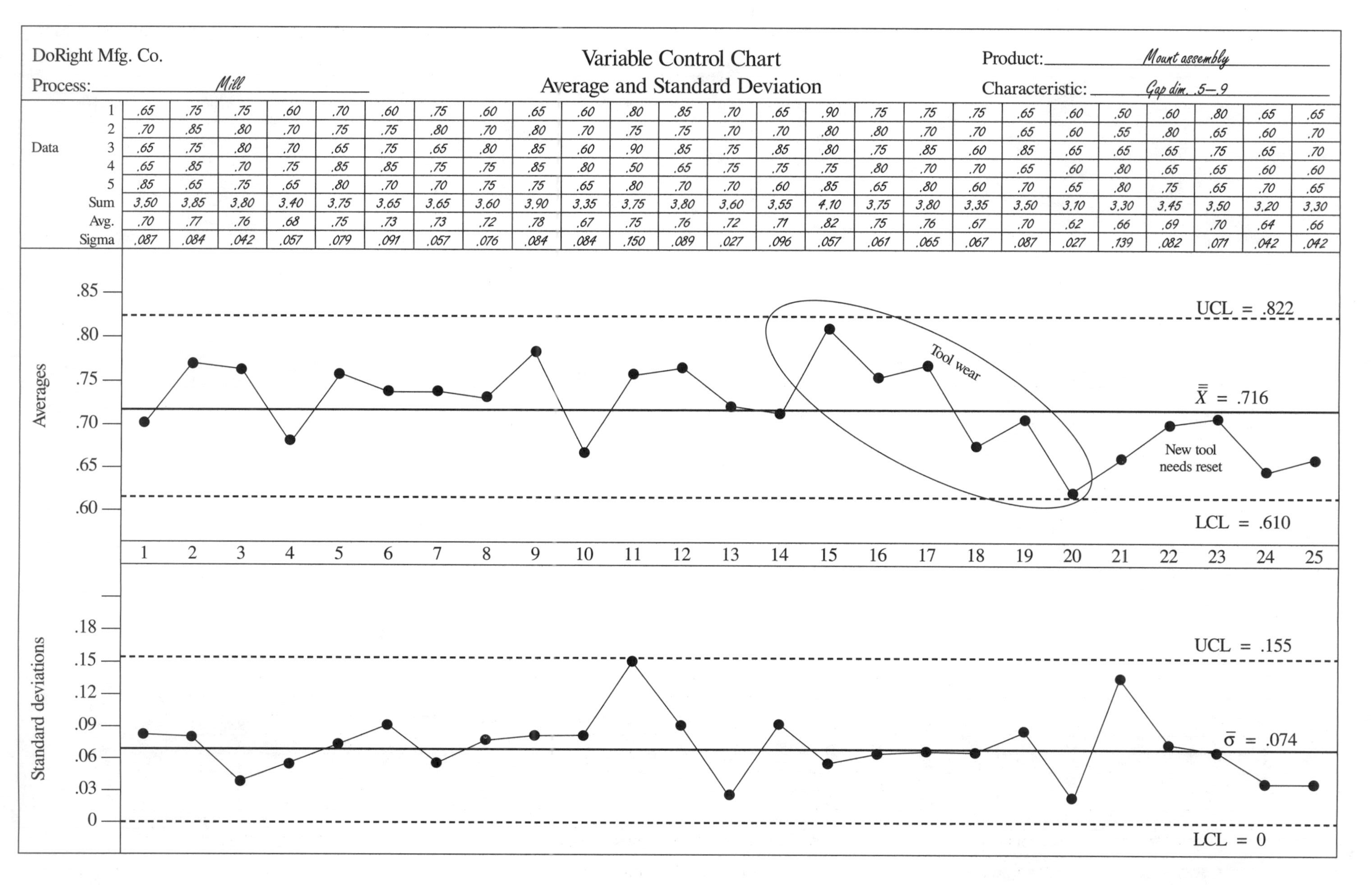

DoRight Mfg. Co.

Process: Mill

Variable Control Chart

Average and Standard Deviation

Product: Mount assembly

Characteristic: Gap dim. .5–.9

		1	2	3	4	5	6	7	8	9	10	11	12	13	14	15	16	17	18	19	20	21	22	23	24	25
Data	1	.65	.75	.75	.60	.70	.60	.75	.60	.65	.60	.80	.85	.70	.65	.90	.75	.75	.75	.65	.60	.50	.60	.80	.65	.65
	2	.70	.85	.80	.70	.75	.75	.80	.70	.80	.70	.75	.75	.70	.70	.80	.80	.70	.70	.65	.60	.55	.80	.65	.60	.70
	3	.65	.75	.80	.70	.65	.75	.65	.80	.85	.60	.90	.85	.75	.85	.80	.75	.85	.60	.85	.65	.65	.65	.75	.65	.70
	4	.65	.85	.70	.75	.85	.85	.75	.75	.85	.80	.50	.65	.75	.75	.75	.80	.70	.70	.65	.60	.80	.65	.65	.60	.60
	5	.85	.65	.75	.65	.80	.70	.70	.75	.75	.65	.80	.70	.70	.60	.85	.65	.80	.60	.70	.65	.80	.75	.65	.70	.65
	Sum	3.50	3.85	3.80	3.40	3.75	3.65	3.65	3.60	3.90	3.35	3.75	3.80	3.60	3.55	4.10	3.75	3.80	3.35	3.50	3.10	3.30	3.45	3.50	3.20	3.30
	Avg.	.70	.77	.76	.68	.75	.73	.73	.72	.78	.67	.75	.76	.72	.71	.82	.75	.76	.67	.70	.62	.66	.69	.70	.64	.66
	Sigma	.087	.084	.042	.057	.079	.091	.057	.076	.084	.084	.150	.089	.027	.096	.057	.061	.065	.067	.087	.027	.139	.082	.071	.042	.042

Figure 2.6. Solution to practice average and standard deviation chart.

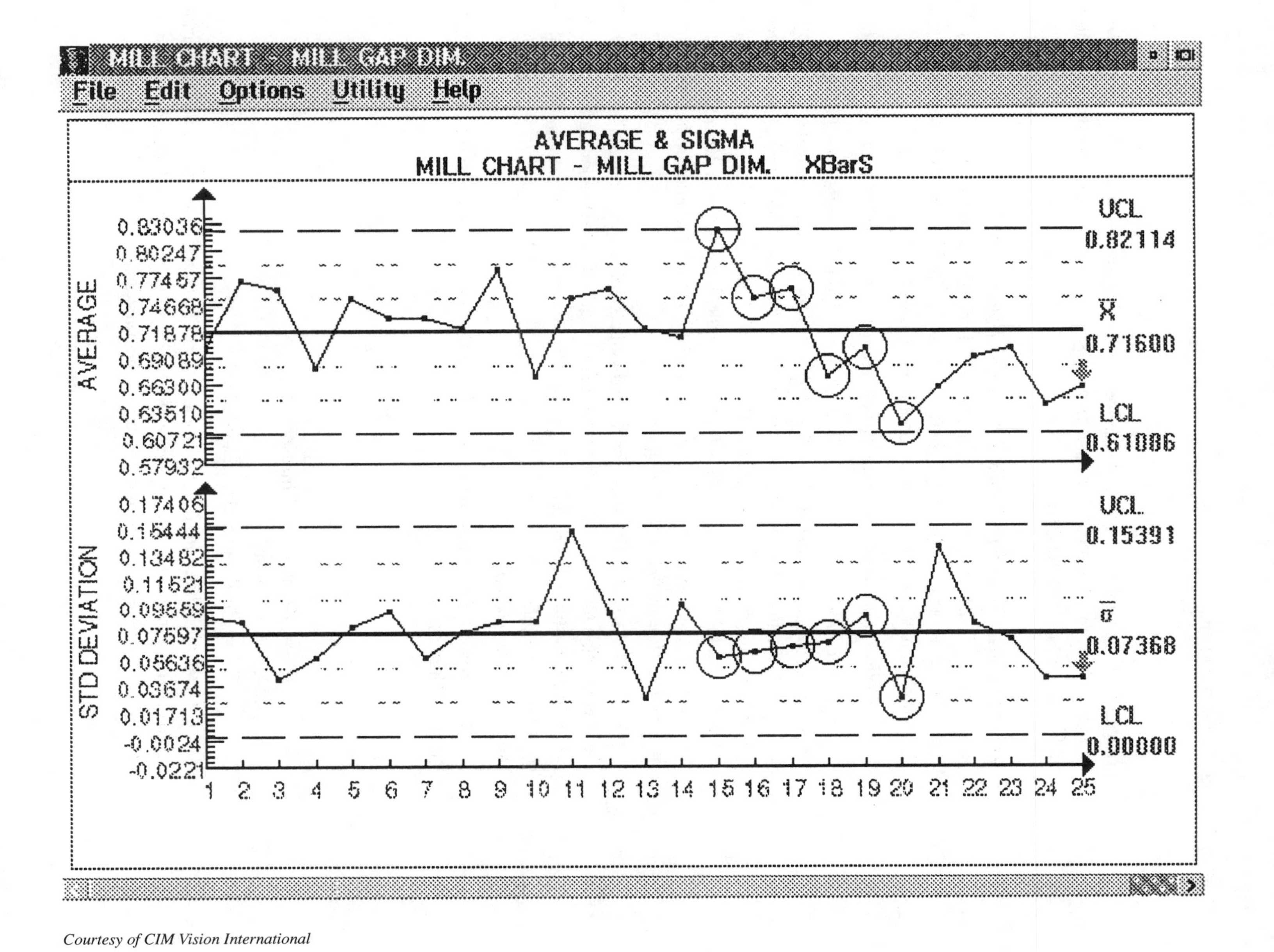

Courtesy of CIM Vision International

Figure 2.7. SQM software example of an average and standard deviation chart.

range charts are not as sensitive to changes in the process as average and range charts. It may be better, in certain cases, to use an average and range chart with small samples and tight intervals to start the chart.

Steps to Construct an Individuals and Moving Range Chart

Refer to Figures 2.8, 2.9, and 2.10 for practice. The control chart form should state that it is an individuals and moving range chart and should reflect the process information such as part number, machine number, data, operation, and so forth.

1. *Collect data:* Individual readings are recorded on the chart from left to right.

2. *Calculate the moving range:* The moving range is the difference between n values each time dropping the first value. For example, say the measurements are 7, 8, 10, 9, 6. If $n = 2$, then the first moving range is calculated using 7 and 8, the second moving range is calculated using 8 and 10, and so on. If $n = 3$, then the first moving range is calculated using 7, 8, and 10. The second moving range is calculated using 8, 10, and 9, and so on. Note: For the purpose of control limits, the sample size used to find the factors is the sample size used to calculate the moving range.

3. *Select the scale:* The scale for the individuals and moving range chart is either: (1) the specification tolerance with additional room for out-of-specification data or (2) one and one-half to two times the difference between the largest and smallest readings. Make sure the scale is identified on the chart. The scale for the moving range chart is the same scale used for the individuals chart.

4. *Determine the centerline of the individuals chart:* Calculate and plot the average of the individual measurements.

$$\bar{X}_i = \frac{\Sigma X_i}{k}$$

where

k = the number of measurements
Σ = the sum
X_i = the individual measurements

5. *Determine the centerline of the moving range chart:* The centerline of the moving range chart is the average moving range ($\bar{R}_m$). The average moving range is the sum of the moving ranges divided by the number of moving ranges.

$$\bar{R}_m = \frac{\Sigma R_m}{k}$$

where

Σ = sum
k = the number of moving ranges
R_m = each moving range

Note: When $n = 2$, there is no moving range for the first measurement so there will be less moving range than there are measurements. When $n = 3$, there are no moving ranges for the first two measurements, and so on.

DoRight Mfg. Co.

Process: Sulphur tank

Variable Control Chart
Individuals and Moving Range

Product: Various

Characteristic: Percent concentration 9–13% $n = 2$

Data	1	2	3	4	5	6	7	8	9	10	11	12	13	14	15	16	17	18	19	20	21	22	23	24	25
1	8.0	8.5	7.4	10.5	9.3	11.1	10.4	10.4	9.0	10.0	11.7	10.3	16.2	11.6	11.5	11.0	12.0	11.0	10.2	10.1	10.5	10.3	11.5	11.1	
2																									
3																									
4																									
5																									
R_m																									

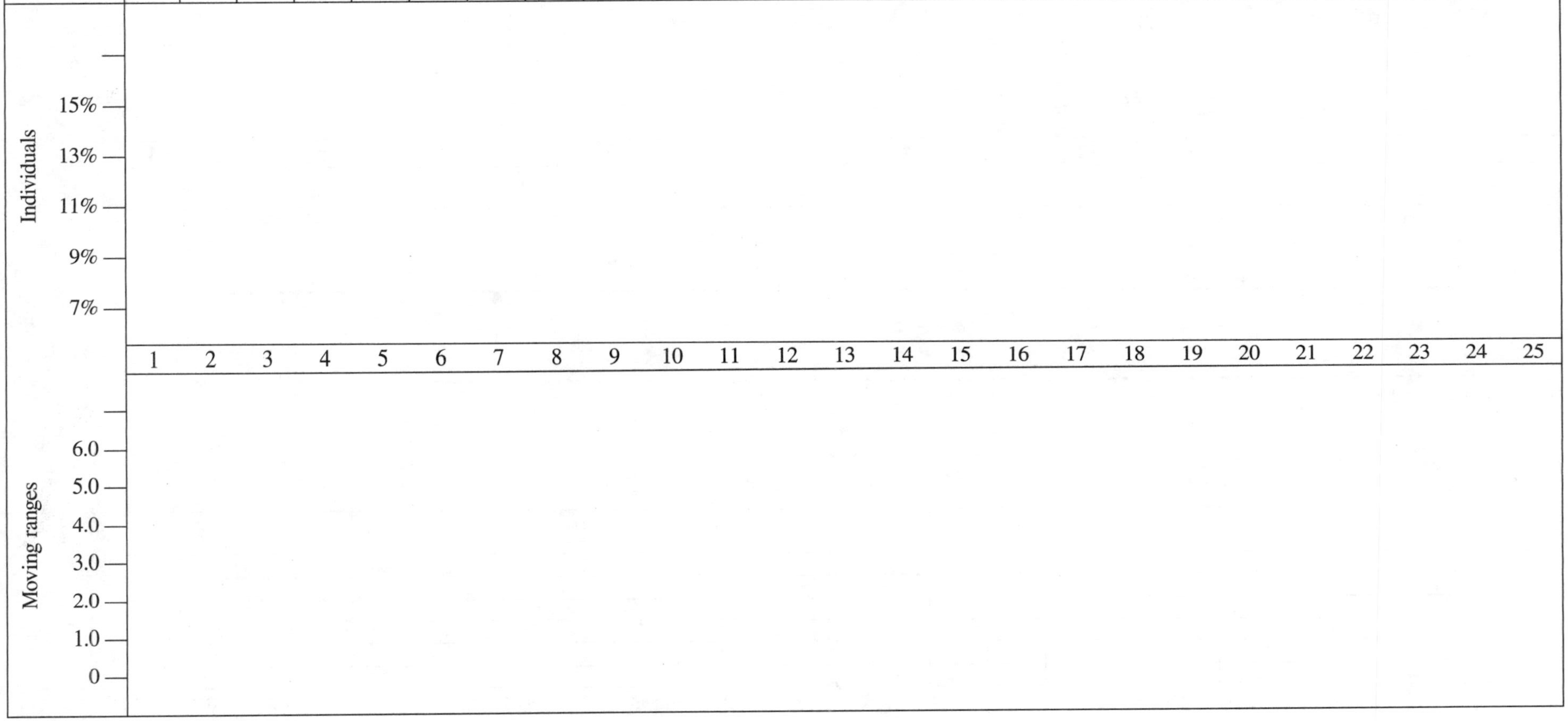

Individuals: 15%, 13%, 11%, 9%, 7%

1 2 3 4 5 6 7 8 9 10 11 12 13 14 15 16 17 18 19 20 21 22 23 24 25

Moving ranges: 6.0, 5.0, 4.0, 3.0, 2.0, 1.0, 0

Figure 2.8. Practice individuals and moving range chart.

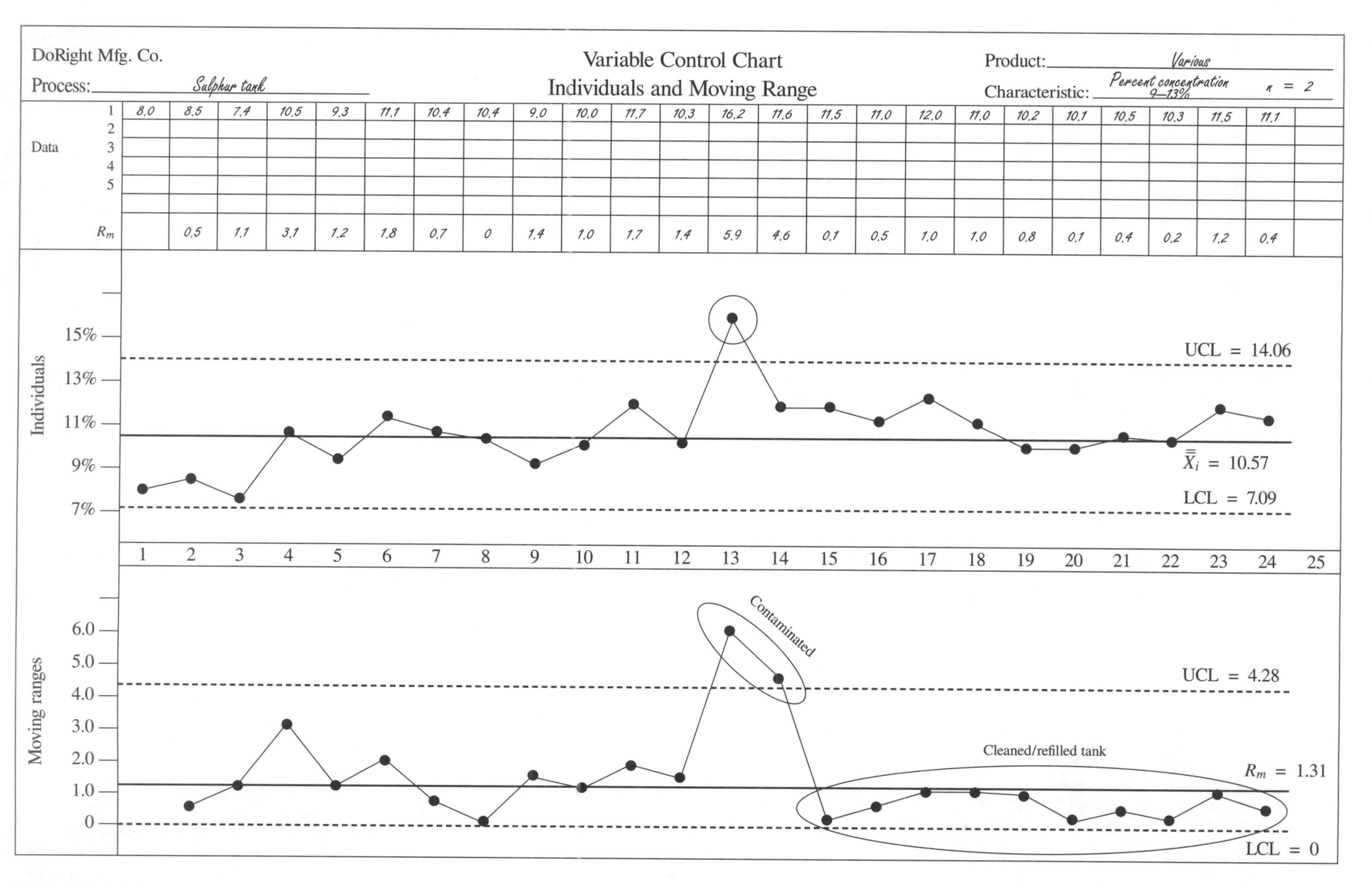

Figure 2.9. Solution to practice individuals and moving range chart.

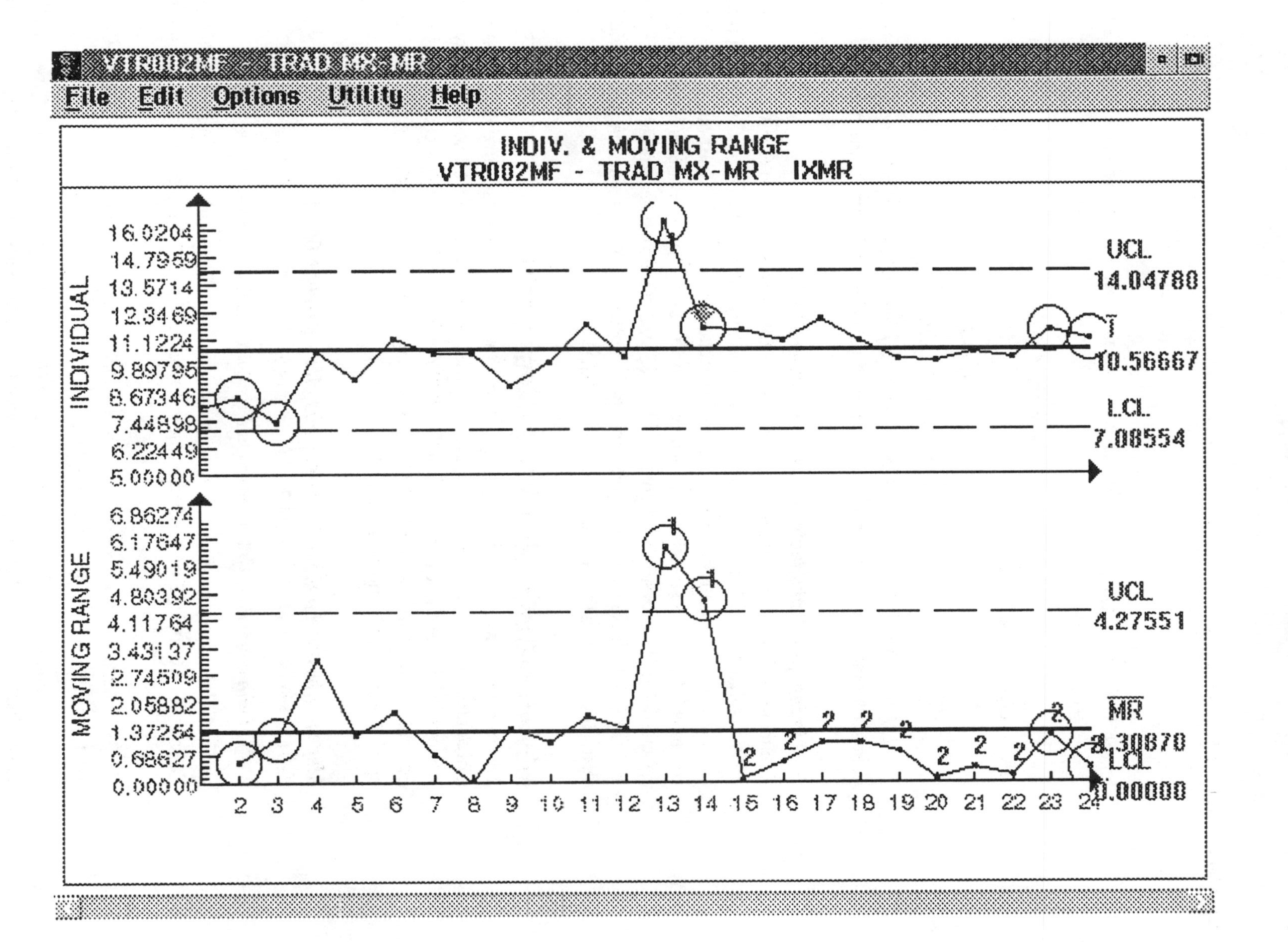

Courtesy of CIM Vision International

Figure 2.10. SQM software example of an individuals and moving range chart.

6. *Calculate the control limits:* The control limits for the charts are calculated as follows:

Individuals chart

$$\overline{X}_i \pm [E_2 \bullet \overline{R}_m]$$

Moving range chart

$$\text{UCL} = D_4 \bullet \overline{R}_m$$

$$\text{LCL} = D_3 \bullet \overline{R}_m$$

7. *Plot:* Plot the individual measurements on the individuals chart and connect the points. Plot the moving ranges on the moving range chart and connect the points.

How to Interpret Individuals and Moving Range Charts

The moving range chart should be reviewed for points that are beyond control limits. These are signs that special causes of variation exist. False trends may occur on the moving range chart because the plotted moving ranges are correlated (they have at least one data point in common). Consult a statistician for advice in these cases. The individuals chart should be reviewed for: (1) points beyond control limits, (2) the spread between points within limits, and (3) trends. Take care when interpreting the spread between points on the individuals chart. It could be large because of a false reading.

MOVING AVERAGE AND MOVING RANGE CHARTS

The moving average and moving range chart ($\overline{X}_m - \overline{R}_m$) is similar in applications to the individuals and moving range chart except it uses a moving average. The moving average and moving range chart is better than the individuals and moving range chart because the moving average dampens some of the effects of overcontrol. It also provides increased ability to detect shifts in the process level.

Steps to Construct a Moving Average and Moving Range Chart

Refer to Figures 2.11, 2.12, and 2.13 for practice.

1. *Data collection:* Collect individual measurements and record left to right.

2. *Calculate the moving average and moving range:* Remember to decide on the sample size; for example, 2 or 3. Then use it to calculate the moving averages and the moving ranges.

3. *Calculate the average ($\overline{X}_m$):* The average is the centerline of the moving average chart. It is the average of all the raw data.

4. *Calculate the average moving range ($\overline{R}_m$):* This is the centerline of the moving range chart. It is the sum of all of the moving ranges divided by the number of moving ranges.

5. *Calculate the control limits for each chart:*

Moving average chart

$$\overline{X} \pm [A_2 \bullet \overline{R}_m]$$

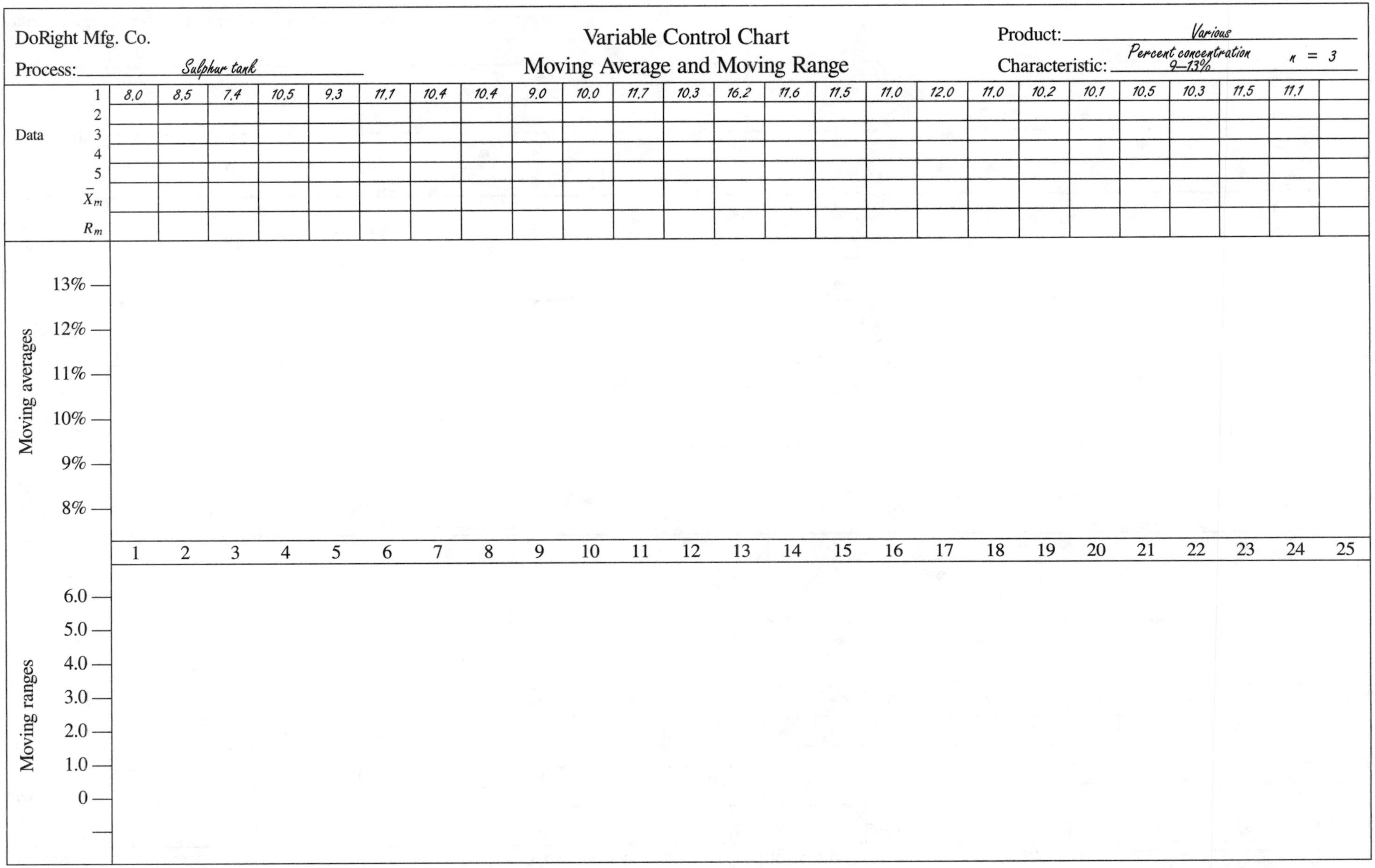

DoRight Mfg. Co.

Process: *Sulphur tank*

Variable Control Chart
Moving Average and Moving Range

Product: *Various*

Characteristic: *Percent concentration 9–13%* $n = 3$

Data	1	2	3	4	5	6	7	8	9	10	11	12	13	14	15	16	17	18	19	20	21	22	23	24	25
1	8.0	8.5	7.4	10.5	9.3	11.1	10.4	10.4	9.0	10.0	11.7	10.3	16.2	11.6	11.5	11.0	12.0	11.0	10.2	10.1	10.5	10.3	11.5	11.1	
2																									
3																									
4																									
5																									
$\bar{X}_m$																									
R_m																									

Moving averages: 13%, 12%, 11%, 10%, 9%, 8%

1 2 3 4 5 6 7 8 9 10 11 12 13 14 15 16 17 18 19 20 21 22 23 24 25

Moving ranges: 6.0, 5.0, 4.0, 3.0, 2.0, 1.0, 0

Figure 2.11. Practice moving average and moving range chart.

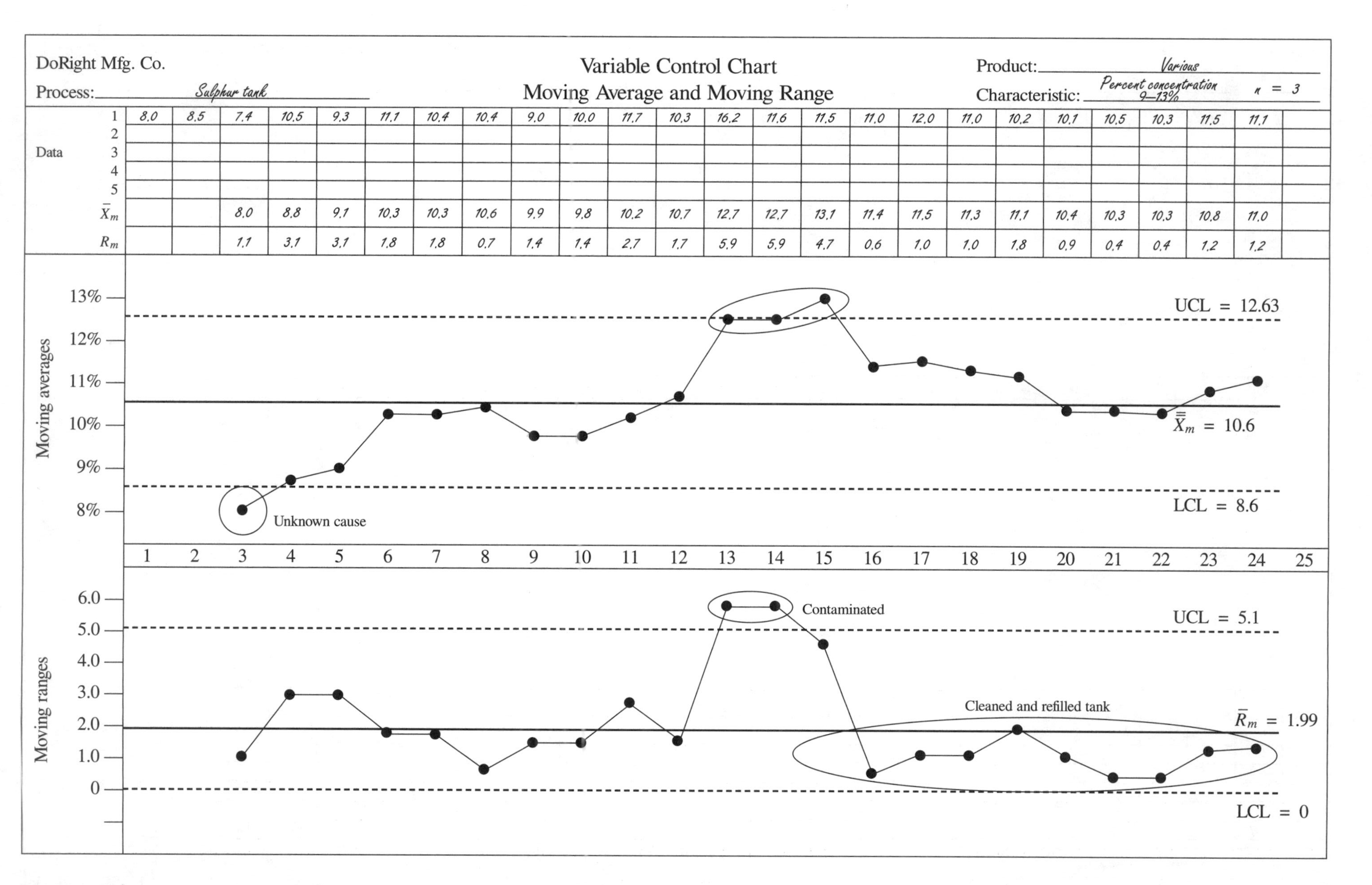

Figure 2.12. Solution to practice moving average and moving range chart.

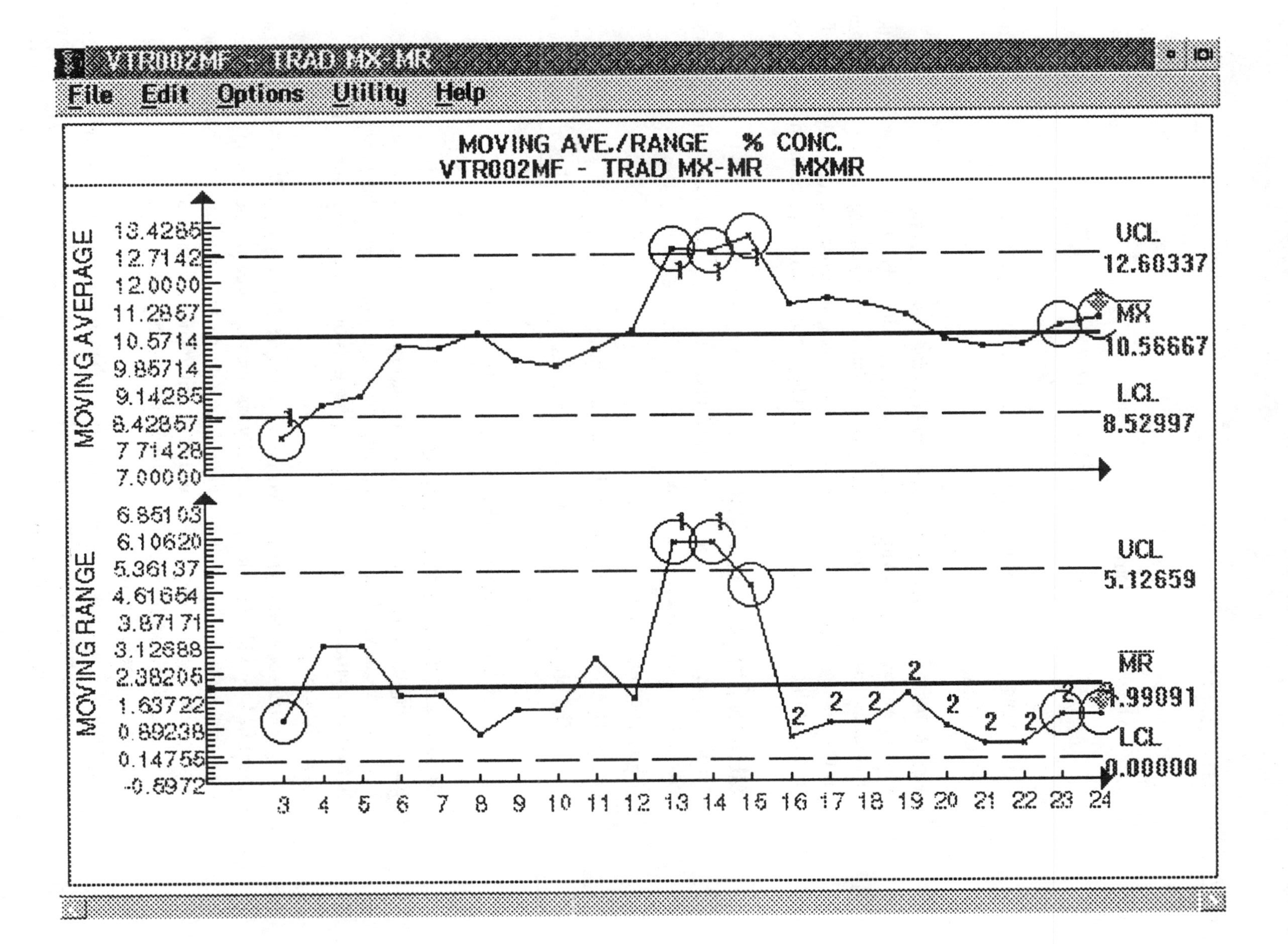

Courtesy of CIM Vision International

Figure 2.13. SQM software example of a moving average and moving range chart.

Moving range chart

$$UCL = D_4 \bullet \bar{R}_m$$
$$LCL = D_3 \bullet \bar{R}_m$$

Note: Factors depend on the sample size used to get the moving average and moving range.

How to Interpret a Moving Averages and Moving Ranges Chart

Moving averages and moving ranges are correlated since they have at least one data point in common. Because moving averages are correlated, one might expect false cycles patterns on the moving average chart. Because moving ranges are correlated, one might expect false trends on the moving range chart. Users should consult with a statistician in these cases.

MEDIAN CONTROL CHARTS

Median control charts ($\tilde{X}$) are another alternative to average and range charts. Some specific advantages of median charts are

1. They are easy to use.
2. They do not require day-to-day calculations.
3. The chart shows the spread of the process output and also gives a picture of the process variation.
4. Users have the option to use a range chart (or not).

Steps to Construct the Medians Chart

Refer to Figures 2.14, 2.15, and 2.16 for practice. It is important to note that medians charts are not as sensitive to variation as average and range charts, but they are very useful to monitor a process that has already had some level of improvements made. The use of the range chart is optional.

1. *Data collection:* Data collection for medians charts is typically 10 parts per subgroup or less. Odd samples (specifically samples of three per subgroup) are preferred because it simplifies finding the median of the data. Refer to the Glossary for calculating the median when sample sizes are even. The discrimination of the measuring equipment, recommended for median charts, should consume only 5 percent of the total part tolerance.

   Example of 5 percent:

   A part has a tolerance of ±.001″ (where the total tolerance is .002″). Five percent of .002″ equals .0001.″ The equipment used to measure the part should discriminate to .0001″ or better.

2. *Scale for the chart:* The scale boundaries for medians charts is typically selected using one of the following two methods. The suggested scale values are equal to the discrimination of the measuring equipment.

   a. The specification tolerance plus extra space on each side for out-of-specification data

   b. One and one-half to two times the range of the measurements

DoRight Mfg. Co.

Process: *Assembly*

Variable Control Chart
Medians and Range

Product: *Shaft/hub assembly*

Characteristic: *Torque 10–18 lb. in.*

n = 3		1	2	3	4	5	6	7	8	9	10	11	12	13	14	15	16	17	18	19	20	21	22	23	24	25
	1	14	15	13	15	16	13	11	14	14	14	11	17	20	18	12	20	17	18	19	16	13	16	18	14	15
	2	12	12	16	13	14	14	17	14	17	17	19	11	19	15	15	14	13	11	13	14	17	14	17	16	16
Data	3	16	16	15	13	17	18	13	17	14	20	15	16	16	13	13	9	15	13	10	13	16	17	13	15	14
	4																									
	5																									
Median																										
Range																										

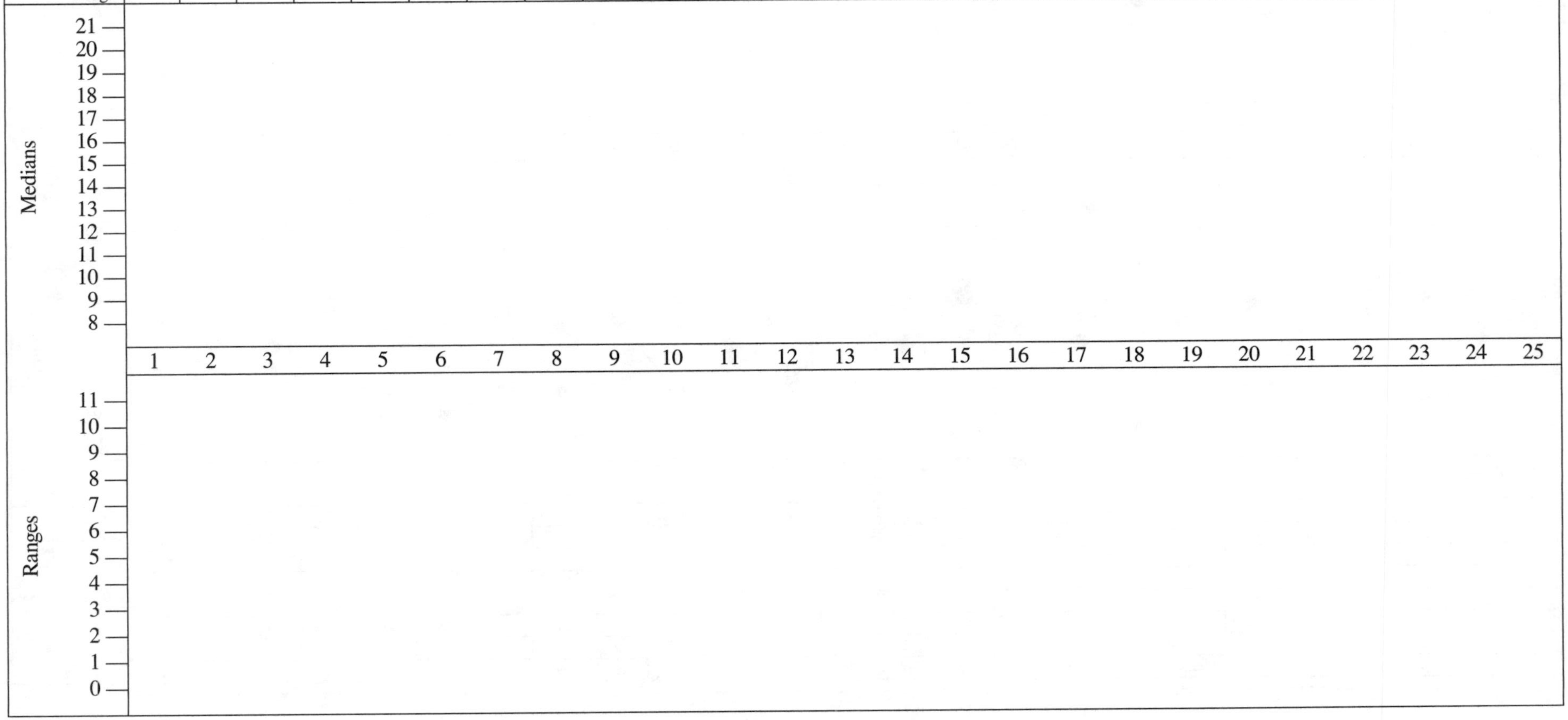

Medians: 21, 20, 19, 18, 17, 16, 15, 14, 13, 12, 11, 10, 9, 8

1 2 3 4 5 6 7 8 9 10 11 12 13 14 15 16 17 18 19 20 21 22 23 24 25

Ranges: 11, 10, 9, 8, 7, 6, 5, 4, 3, 2, 1, 0

Figure 2.14. Practice medians and range chart.

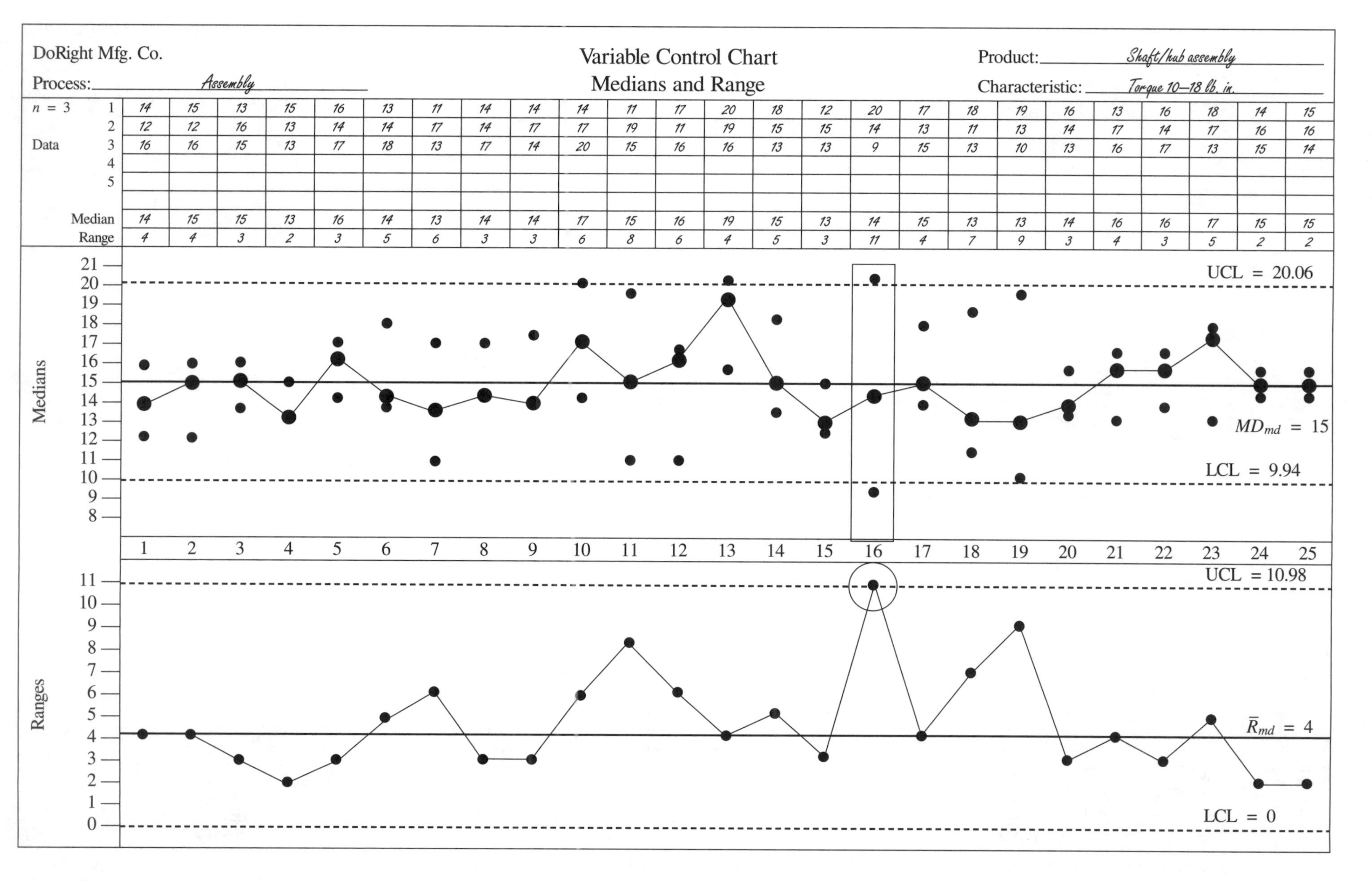

DoRight Mfg. Co.

Process: *Assembly*

Variable Control Chart
Medians and Range

Product: *Shaft/hub assembly*

Characteristic: *Torque 10–18 lb. in.*

n = 3																									
Data 1	14	15	13	15	16	13	11	14	14	14	11	17	20	18	12	20	17	18	19	16	13	16	18	14	15
2	12	12	16	13	14	14	17	14	17	17	19	11	19	15	15	14	13	11	13	14	17	14	17	16	16
3	16	16	15	13	17	18	13	17	14	20	15	16	16	13	13	9	15	13	10	13	16	17	13	15	14
4																									
5																									
Median	14	15	15	13	16	14	13	14	14	17	15	16	19	15	13	14	15	13	13	14	16	16	17	15	15
Range	4	4	3	2	3	5	6	3	3	6	8	6	4	5	3	11	4	7	9	3	4	3	5	2	2

Figure 2.15. Solution to practice medians and range charts.

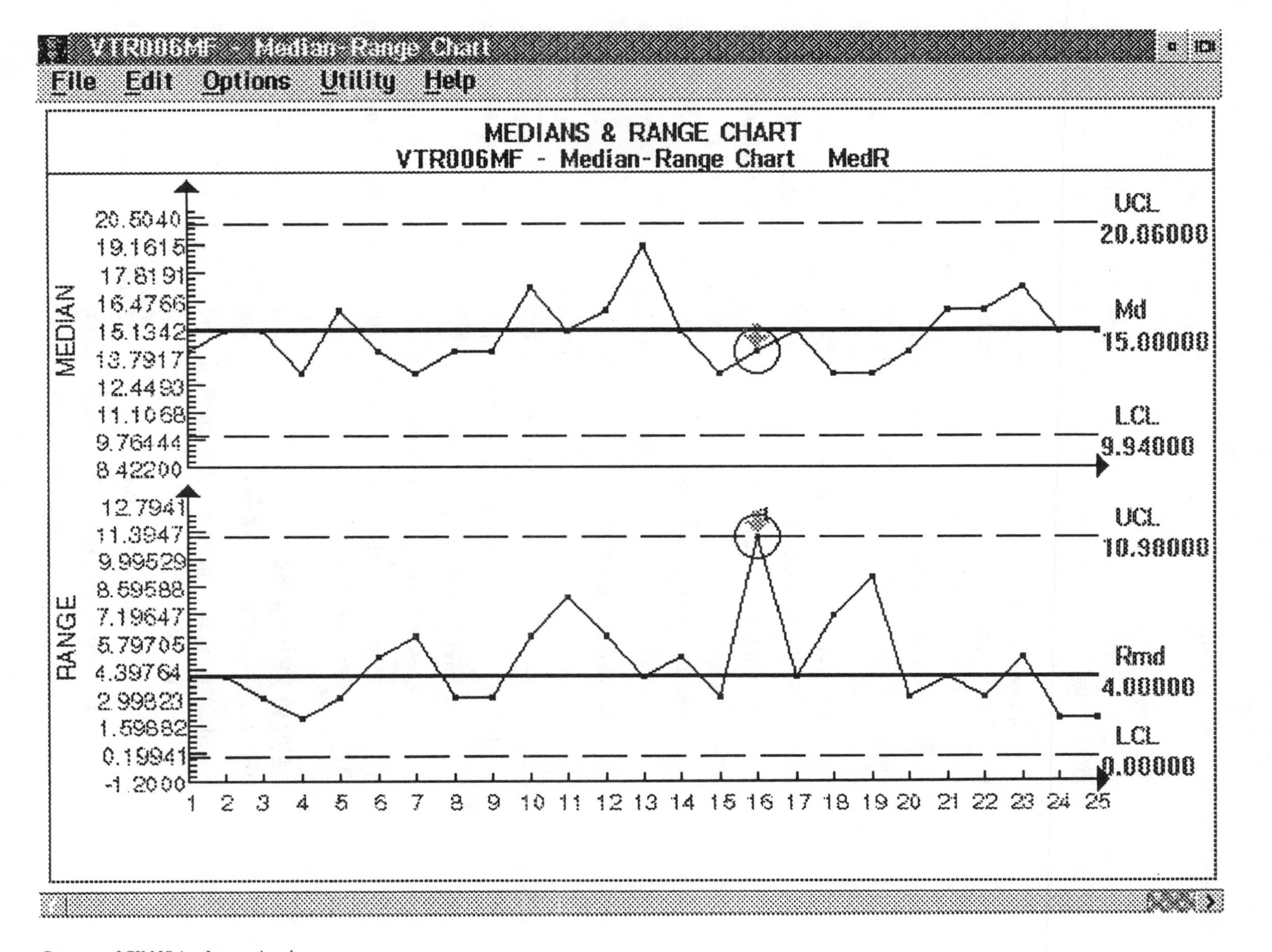

Courtesy of CIM Vision International

Figure 2.16. SQM software example of medians and range chart. Note that only the medians are plotted because the range chart is optional.

3. *Medians and ranges:* Enter the median (*Md*) and range (*R*) values for each subgroup in the data block. Note: For this example, I prefer to use an accompanying range chart.

4. *Plotting:* All of the measurements in the sample are plotted on the same vertical line, and the value of the median is usually circled. Then, all of the medians from left to right are connected using lines.

5. *The grand median (Md_{Md}):* The grand median is the median of all subgroup medians that are arranged in ascending or descending order. It is the centerline of the medians chart.

6. *The median range (R_{md}):* The median range is the median of all subgroup ranges that are arranged in ascending or descending order. It is the centerline of the range chart.

7. *Control limits:* Calculate the control limits for both charts. Refer to Appendix B for the A_5, D_6, and D_5 factors.

Medians chart control limits

$$Md_{Md} \pm A_5 \bullet R_{Md}$$

Range chart control limits

$$UCL_R = D_6 \bullet R_{Md}$$

$$LCL_R = D_5 \bullet R_{Md}$$

8. *Plotting:* Plot the centerline and control limits on the median and the range chart.

9. *Interpretation:* Interpret the charts.

   a. Look for any median value that is beyond the median control limits.

   b. About two-thirds of the median values should fall within the middle one-third area between the control limits.

REVIEW PROBLEMS

Refer to Appendix H for answers.

1. Arrange the following measurements in a frequency distribution.

   .60 .80 .70 .75 .65 .70 .65 .70

2. Is the frequency distribution in problem 1 normal shaped? (Yes or no)

3. Find the average ($\overline{X}$) of the following measurements.

   .500 .501 .496 .502 .499

4. Find the range (R) of the following measurements.

   .501 .500 .496 .502 .499

5. Find the grand average ($\overline{\overline{X}}$) of the following averages.

   50 72 57 48 53 53 70 69 67 59

6. Find the average range ($\overline{R}$) of the following ranges.

   6 7 9 5 9 6 7 8 7 6

7. Find the UCL for an averages chart using the following information.

   $n = 5 \quad \bar{\bar{X}} = 50 \quad \bar{R} = 2$

8. Find the UCL for a range chart using the following information.

   $n = 4 \quad \bar{R} = 8$

9. Find the LCL for a range chart using the following information.

   $n = 5 \quad \bar{R} = .002$

10. Find the estimated standard deviation for control charts using the following information.

    $n = 5 \quad \bar{R} = 10$

11. A process can be in statistical control and still make products out of specification limits. (True or false)

12. Variation that lies within the normal distribution is due to common causes. (True or false)

13. The first objective of SPC is to get a process under statistical control before studying its capability. (True or false)

14. A process is

    a. A machine

    b. The tooling

    c. The operator

    d. All of the above and more

15. Special causes of variation exist in a process when

    a. Parts are out of specification

    b. Points are within control limits

    c. Parts are all to print

    d. Points are outside control limits

16. How many total standard deviations around the mean are equal to 99.73 percent of the normal distribution?

    a. 3

    b. 2

    c. 6

    d. 1

17. Things that can be measured are called

    a. Attributes

    b. Nominals

    c. Variables

    d. None of the above

18. Find the word below that closely matches the meaning of *statistical control*.

    a. Ability

    b. Repeatable

    c. Accurate

    d. Close

19. Find the first moving range of the following data where $n = 3$.

    5 6 8 5 9

20. Calculate the average moving range for these moving ranges.

    .002 .003 .002 .003 .001

21. Find the UCL for an individuals chart using the following information.

    $\bar{X} = .500$ $\bar{R}_m = .001$ $n = 3$

22. Find the UCL for a moving range chart using the following information.

    $\bar{R}_m = .001$ $n = 3$

23. Find the estimated standard deviation using the following information.

    $\bar{R}_m = .001$ $n = 3$

24. Find the first moving average for the following data where $n = 2$.

    .500 .502 .503 .499 .501

25. Find the median of the following data.

    12 17 10 8 8 11 14

26. What percentage of the part tolerance is recommended for selecting the gage for a median chart?

27. Find the UCL for a medians chart using the following information.

    $MD_{MD} = .498$ $R_{Md} = .003$ $n = 5$

CHAPTER 3

Short-Run Variables Control Charts

INTRODUCTION

Statistical process control for moderate and short production runs is a matter of knowledge of control charting and following some short-run rules. In fact, the rules to follow are the rules of SPC regardless of the length of production runs. SPC is often misinterpreted as a technique that controls the products' characteristics or dimensions when, in fact, the methods were originally intended to control the process with the understanding that a controlled process will produce products that are more consistent.

Process control is concerned primarily with the repeatability of the process. Much of the power of control charts is wasted when the stated purpose is to control a single characteristic. One should perform the necessary brainstorming to pick product characteristics that can be "used" for process control that will focus on the process. There are cases in which certain characteristics of a product are important for several reasons such as, fit, form, function, and failure quality costs. In these cases, these characteristics should be taken into consideration for control in such a manner that they may be the characteristic measured or inherently controlled by the chart(s). When applying SPC, always try to use techniques that will minimize the amount of charts and, at the same time, maximize the effect. For example, a single control chart is constructed using a characteristic that focuses on the process and helps to eliminate special causes that adversely affect several characteristics of the product.

There are many benefits that can be obtained by using short-run SPC methods. The following is a brief list.

1. Significantly fewer charts to be maintained.
2. Process variability improvements affect many part numbers, not just one.
3. Process control is continuous (one chart for many parts).
4. Control methods are focused on the process, not the part number.
5. Individual part number information can be obtained when needed.

DEFINITION OF A SHORT RUN

A short-run problem can be characterized in several ways, but the problem always narrows down to insufficient or untimely data for control limits. When asked, people from different companies and different product lines have different perceptions of a short run. For example, a toothpick manufacturer may call a run of 50,000 toothpicks a short run, while a casting manufacturer may feel that 400 castings is a short run. In many aerospace companies, short runs have been identified as ranging from 1 to 12 parts.

Short-run SPC problems lie in the following general categories.

1. Not enough parts in a single production run to achieve/maintain control limits on the process.
2. The process cycles so quickly that even large size production runs are over before data can be gathered.
3. Many different parts are made for many different customers (in small lot sizes).

The first lesson learned in short-run SPC is that SPC is not about parts, it's about the process. Parts come and go, but the process is, more or less, continuous.

RULES FOR SHORT-RUN SPC

The following rules are important in solving the problem of statistical process control in short production run applications and, in many cases, also prove effective in some mass production applications. When applied properly, these rules help optimize the process control, reduce the amount of charts required, and increase the benefits obtained from SPC. These rules are

Rule 1. Focus your attention on the process. The objectives of SPC are to improve products made by the process. SPC methods, however, must focus on the process itself. Monitoring and controlling the variation of the process is required in order to improve the product. SPC methods are about the process, not the parts.

Rule 2. It must be the same process stream. The method should be applied to one process at a time because every process has its own inherent variation. Confusing more than one process on the same chart will give false signals about process control. The only exception to this rule is with the use of standardized control charts, which are covered later in this chapter.

The method of processing should also be the same due to the fact that different processing methods (on one process) can also be the source of false signals.

In many processes, the type of material should also be a constant. Different materials vary differently on one process. For example, aluminum varies differently than steel on a lathe. Refer to rule 6 about representative variation.

Rule 3. Look for generic families of product made by the process. This rule works with the others to help ensure that there is a lot in common among several part numbers that are to be charted together on one chart. For example, one commonality is covered by rule 2 (which calls for a material family) or parts made from the same (or very similar) materials.

Rule 4. Use coded data. Coded data are the reason why different parts with different target dimensions and tolerances can be charted together. Using coded data makes all charts robust to different part numbers, nominal dimensions, and tolerances, therefore, coded data make all traditional control charts become short-run control charts.

Rule 5. Statistical control charts require a minimum of 20 subgroups of data (not parts). This basic rule is standard for all statistical control charts. Using short-run charts and coded data, the minimum subgroups of data can more easily be obtained.

Rule 6. Variation between different part numbers charted on the same chart must be representative (for example, they vary similarly). Using coded data helps when charting several different part numbers on the same chart, but if their variation is significantly different, false signals can occur. The different parts must also be a representative family of variability. The only exception is covered later in this chapter.

Rule 7. Measurements should be made to a fine enough degree to detect variation. Some basic rules of measurement are involved such as: (1) The measuring instrument should discriminate to no more than 10 percent of the part tolerance, and (2) The repeatability and reproducibility (gage R&R) of the measurement system (the observer and the gage) should not exceed a certain percentage of part total tolerance (see chapter 8).

If all of these rules are followed during the planning for SPC, many problems will be overcome before they happen.

PRODUCT FAMILIES FOR SHORT-RUN SPC

One of the rules for short-run SPC is to identify product families. There are a variety of families from which to choose, as shown in Table 3.1. The important thing is to look at process control as a tool to control the entire family, not just one part number within the family.

Short-run SPC methods allow different part numbers, targets, and specifications to be charted on one process-focused chart, so these differences in the parts are not a problem with a short-run chart.

A family is merely products that are made by the same process that have common traits such as the same material, configuration, or type of control characteristic. Knowing what the different part numbers have in common and what they do not have in common is the basis for applying short-run methods. The examples of families in Table 3.1 are food for thought toward understanding short-run SPC applications.

STATISTICS USING CODED DATA

Short-run SPC methods work on a variety of different part numbers, target dimensions, and part tolerances primarily due to the use of coded data. There are basically two methods for coding the measurements (data) from a process. These coding methods are as follows:

- *Deviation from target* is a coding method in which measurements are made on the product (or process variable), but only the difference between the measured value and the target value is recorded. For example, if the target of the specification is .500, and the measurements are accurate to .001," actual measurements of .502 and .497 would be recorded as +.002 and –.003, respectively.

As another example, say three part numbers are in a family of parts to be charted on the process for a specific outside diameter. The nominal dimensions and tolerances for these

Table 3.1. Short-run SPC applications.

Family	Process example	Similarities	Differences	Key characteristics	Chart/ remarks
Part	Extrusion	All tubes, all aluminum, same tolerance	Nominal dimensions	Outside diameter	Target average and range chart
Part	Press	Same part type, same tolerance	3 different materials	Height	Target average and range chart/one for each material
Spindle	Drill press	Drilled holes, same material	4 spindles, different sizes and tolerances	Inside diameter	Target average and range chart/ one for each spindle (or a group target chart)
Parameter	Heat treat	Same process	Different parts, and hardness requirements	Temperature	Target average and range chart
Fast	High-speed roll forming	All aluminum fins, same tolerance	Nominals	Height	Target average and range chart
Characteristic	Mill	All steel, same characteristics, same tolerances	Nominal sizes	Depth	Target average and range chart
Tool/cutter path	N/C lathe	All are diameters, cut by the same tool, in the same machine axis, using the same N/C program	Nominals and tolerances	Diameter	Target average and range chart (on the largest diameter furthest from the chuck)

parts are shown in Table 3.2. Measurements are made on each of the part numbers and they are coded to where zero equals the nominal dimension. The results are shown in Table 3.3.

Coding data using the deviation from target method is easy. All that is needed is a base (0 = nominal) and measurement accuracy (.001″). With regard to statistical control charts, the tolerance on each part has no effect on statistical limits. When it comes time to study the capability of the process, however, capability must be studied with respect to each part tolerance.

• *Standardized plot points* is another, more complex, coding method that factors the plot points on a chart based on their history. Refer to standardized control charts covered later in this chapter for more information.

At this point, the deviation from target method of coding data can, and does, make any traditional control chart a short-run chart.

Table 3.2. Table of part numbers.

Part number	Nominal	Tolerance
20224-1	.750	±.020
34520-2	.250	±.010
48028-4	.500	±.005

Table 3.3. Table of coded measurements.

Part number	Nominal	Actual measurement	Coded measurement
20224-1	.750	.753	+.003
34520-2	.250	.249	–.001
48028-4	.500	.500	0

With short-run control charts that use coded data, the equations for computing the control limits are the same, except computations can become more difficult due to negative numbers in the data. For this reason, one must follow basic algebraic rules for adding, subtracting, multiplying, and dividing data that contains negative numbers. Table 3.4 shows the basic statistics computed where the raw data have negative numbers.

SHORT-RUN TARGET CONTROL CHARTS

Target charts are those that use coded data where the basis for coding the data is the specification target or the historical target of the process. The coding of data for these charts consists of just recording the deviation from target values in accordance with the discrimination of the measuring instrument (for example, +.001 or –.002, and so on).

There are a variety of different specification targets depending on the type of specification. The following is a set of examples of specifications and their associated targets.

Table 3.4. Table of statistics using negative values.

Statistic	Data	Solution	Answer
Sum (Σ)	+2, 0, –3, +5, +1	2 + 0 = 2 . . . + –3 = –1 . . . +5 = 4 . . . +1 = 5	5
Average ($\overline{X}$)	+2, 0, –3, +5, +1	5/5 = 1	1
Range (*R*)	+2, 0, –3, +5, +1	High +5 minus low –3	8
Standard deviation (*s*)	+2, 0, –3, +5, +1	Calculator (changing negative numbers to minus before entering them into memory).	2.92

Specification type	Example	Target value example
Bilateral	.500 ± .005	.500 (nominal)
Unilateral	.750 maximum	Any desired value (for example, .600)
Position tolerance	1.250 basic	1.250 basic

Historical targets are less involved with specifications and more involved with the actual target value of the process. A *historical target* is defined as the actual value of the average output of the process. A specification may have a nominal value of .500, but the process has always been targeted at .502 (and .502 is an acceptable target for the output).

Historical targets are used instead of specification in the following general cases.

1. The process is not easily adjusted.
2. The historical target value is preferred over the specification target value.
3. A single limit specification (maximum or minimum) where high is best.
4. A single limit specification (maximum or minimum) where low is best.
5. There are existing data that can be used as history to start the chart.

Examples of historical target values and their applications are shown in Table 3.5.

TARGET AVERAGE AND RANGE CHARTS (DESIRED TARGET)

The traditional average and range chart becomes a short-run control chart when coded data are used, and the centerlines and control limits are computed in the same manner discussed in chapter 2, except that coded data are used. Typical applications for the target average and range chart are

1. Adjustable processes
2. A desired target exists (for example, nominal or basic dimensions)
3. The tolerance for the parts is bilateral (unless it is a single limit tolerance where a desired value has been derived.).

A desired target indicates that the processor will be attempting to make adjustments toward that target during processing (hence the process must be one that is adjustable by the operator).

Table 3.5. Historical targets examples.

Specification target	Historical target	Remarks
.500 nominal dimension	.502	Process is not easily adjusted, and .502 is an acceptable output target for the product.
63 maximum surface finish	30 finish	The output described as low is best, and the existing output finish (30) is being obtained without extra effort.
.100 minimum wall thickness	.175 wall	The output described as high best, and the existing output wall (.175) is being obtained without extra effort.

The setup of the chart uses the same steps as in the average and range chart, except (1) Coded data are used, (2) Notes are taken that identify subgroups where a new part number is set up, (3) Tests for representative variability are conducted, and (4) Capability studies involve studying more than one tolerance. Refer to Figures 3.1, 3.2, and 3.3 for practice in setting up and plotting the desired target average and range chart.

HISTORICAL TARGET AVERAGE AND RANGE CHARTS

Typical applications for the historical target average and range chart are

1. Processes that are not adjustable, such as forging, casting, roll forming, blanking, punching, or drilling.
2. Products that have no nominal or basic dimension, such as geometric tolerances, surface finish, or maximum or minimum limit dimensions.
3. Products that should not be centered on specification tolerances such as (1) High is best, (2) Low is best, or (3) Other is best.

A *best target value* can be described as that value where the product or process will work best. A few best targets and examples follow.

• *High is best* is where a product characteristic or process parameter will work best at or toward the high side of the specification (and sometimes above the upper specification limit where future processing will finish that characteristic). An example could be a hole that is held on the high side so that a future plating process will bring the hole size into specifications. It is understood, in this example, that the dimension of the hole applies after plating.

• *Low is best* is where a product characteristic or process parameter will work best at or toward the low side of the specification (and sometimes below the lower specification limit where future processing will finish that characteristic). An example could be the surface finish of a part, which, in most cases, is a better part.

• *Other is best* is an exact value, within specification limits, that is desirable for functional or cost reasons. For example, a process is not adjustable (without considerable expense and effort), and the C_{pk} index is very good. It may be desired to leave the process at the existing target and control it around that target.

A historical target may also be used where the control limits of a process are calculated based on historical (existing) data. When using historical target charts, the basis for coding the data for the averages chart is the historical average of the part. For example, if surface finish is being charted, the historical grand average ($\overline{\overline{X}}_H$) is the base for coding surface finish measurements. In this case, however, the historical grand average of each different part number must be known. The basis for coding data for each part is the grand average of each part.

Steps to Construct the Historical Target Average and Range Chart

Refer to Figures 3.4, 3.5, and 3.6 for practice setting up and plotting a historical target average and range chart.

DoRight Mfg. Co.

Process: *Wax mold*

Variable Control Chart
Short-Run Target Average and Range

Product: *Covers*

Characteristic: *O/A length*

	1	2	3	4	5	6	7	8	9	10	11	12	13	14	15	16	17	18	19	20	21	22	23	24	25
Data 1	.001	.002	.003	.001	.001	−.001	0	.002	.002	.004															
Data 2	0	.001	.001	−.001	−.002	−.003	0	−.002	0	−.002															
Data 3																									
Part #	1	1	1	2	2	2	1	3	3	3															
Sum	.001	.003	.004	0	−.001	−.004	0	0	.002	.002															
Avg.	.0005	.0015	.002	0	−.0005	−.002	0	0	.001	.001															
Range	.001	.001	.002	.002	.003	.002	0	.004	.002	.006															

+.005
+.004
+.003
+.002
+.001
0
−.001
−.002
−.003
−.004
−.005

Note: This chart is for example only. More subgroups required to derive control limits.

Target = Nominal

1 2 3 4 5 6 7 8 9 10 11 12 13 14 15 16 17 18 19 20 21 22 23 24 25

.008
.007
.006
.005
.004
.003
.002
.001
0

Cover part numbers		
Part #	Nominal	Tolerance
2203-5	.500	±.003
30517-9	.750	±.005
13156-1	.375	±.010

Vertical line indicates new part number setup.

Figure 3.1. Practice short-run target chart. Nominal = 0.

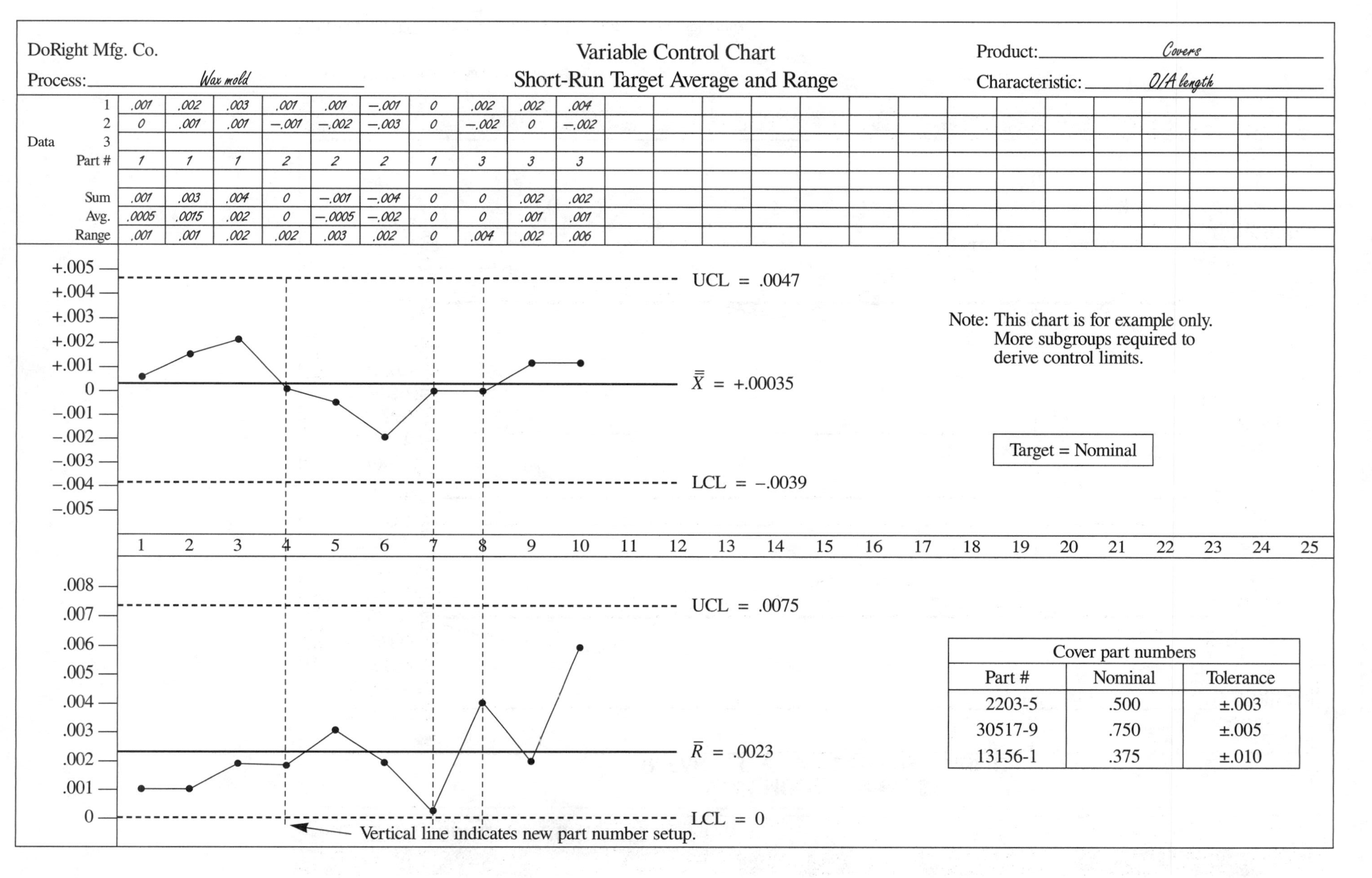

	1	2	3	4	5	6	7	8	9	10
Data 1	.001	.002	.003	.001	.001	–.001	0	.002	.002	.004
Data 2	0	.001	.001	–.001	–.002	–.003	0	–.002	0	–.002
Data 3										
Part #	1	1	1	2	2	2	1	3	3	3
Sum	.001	.003	.004	0	–.001	–.004	0	0	.002	.002
Avg.	.0005	.0015	.002	0	–.0005	–.002	0	0	.001	.001
Range	.001	.001	.002	.002	.003	.002	0	.004	.002	.006

Cover part numbers		
Part #	Nominal	Tolerance
2203-5	.500	±.003
30517-9	.750	±.005
13156-1	.375	±.010

Figure 3.2. Solution to practice short-run target chart.

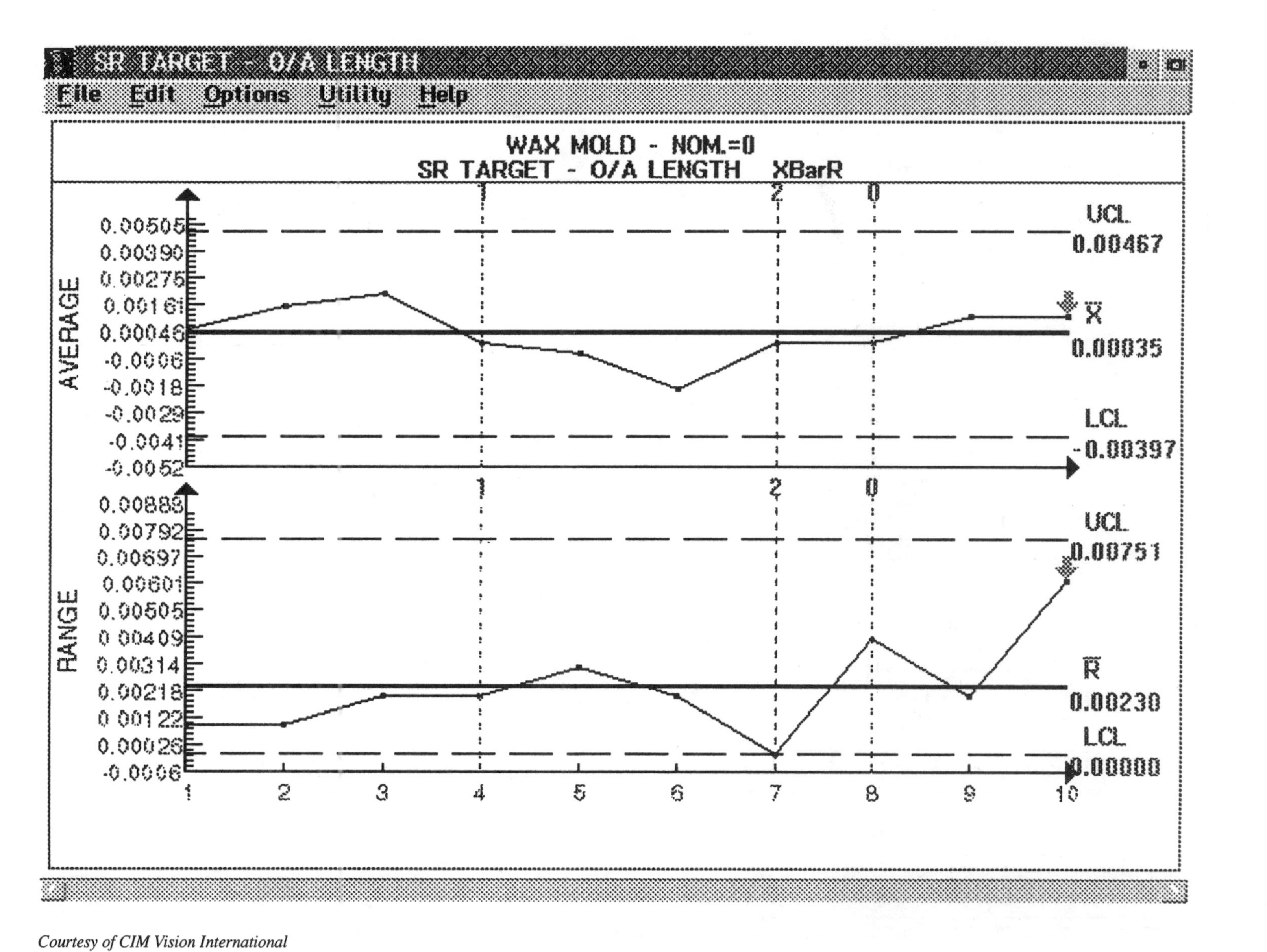

Courtesy of CIM Vision International

Figure 3.3. SQM software example of a short-run target chart.

DoRight Mfg. Co. $n = 2$

Process: *Lathe*

Variable Control Chart
Historical Target Average and Range

Product: *Housings*

Characteristic: *Surface finish*

	Part #	A	A	B	B	C	C	D	A					
Data	1	57	59	28	31	24	22	36	61					
	2	59	60	30	29	21	20	42	58					
	Sum	116	119	58	60	45	42	78	119					
	$\bar{X}$													
	Range													
	$\bar{\bar{X}}_H$	60	60	30	30	24	24	40	60					
	$\bar{X}-\bar{\bar{X}}_H$													
		1	2	3	4	5	6	7	8	9	10	11	12	13

Note: The historical average for each part need not be updated if the average of the part is in control. Constant updating, however, is a user option.

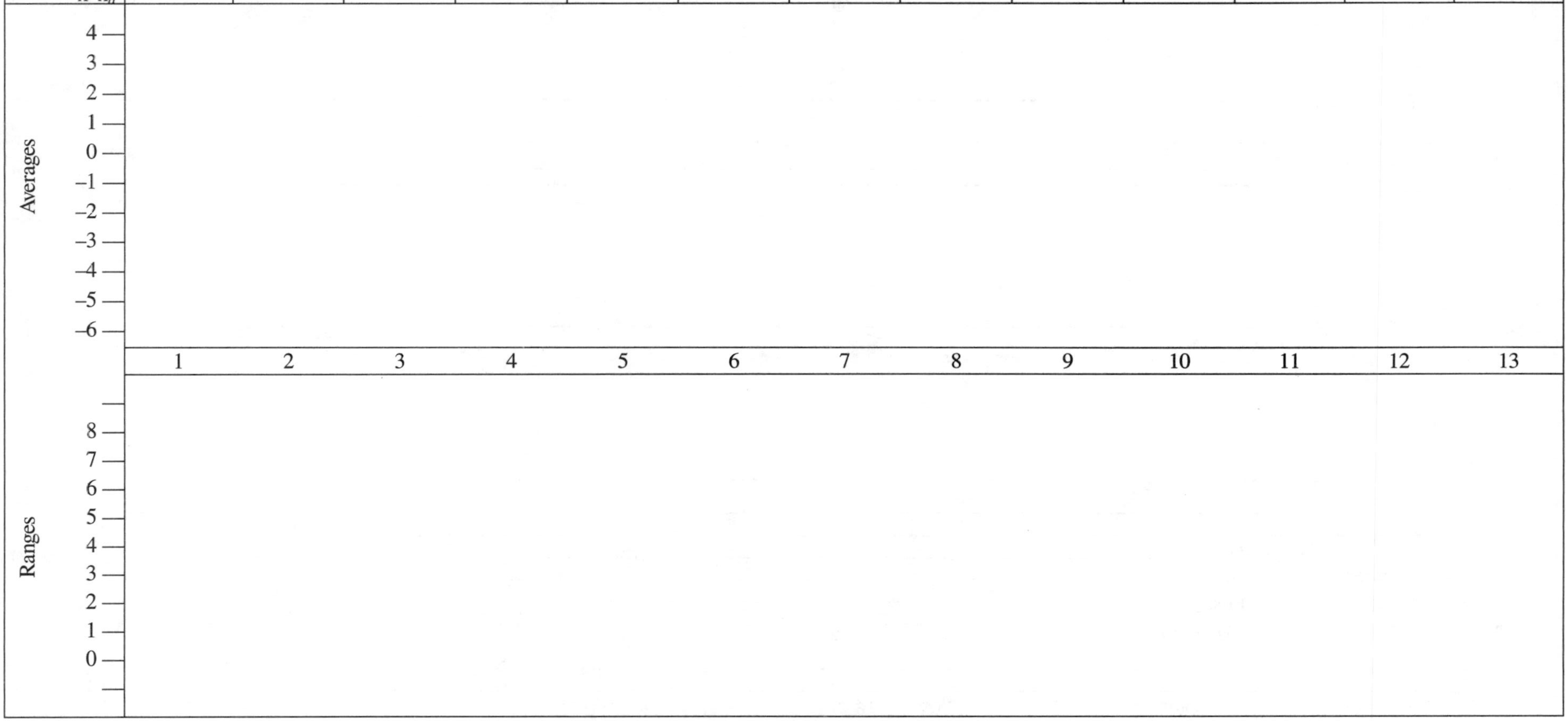

Averages: 4, 3, 2, 1, 0, –1, –2, –3, –4, –5, –6

Ranges: 8, 7, 6, 5, 4, 3, 2, 1, 0

Figure 3.4. Practice historical target average and range chart.

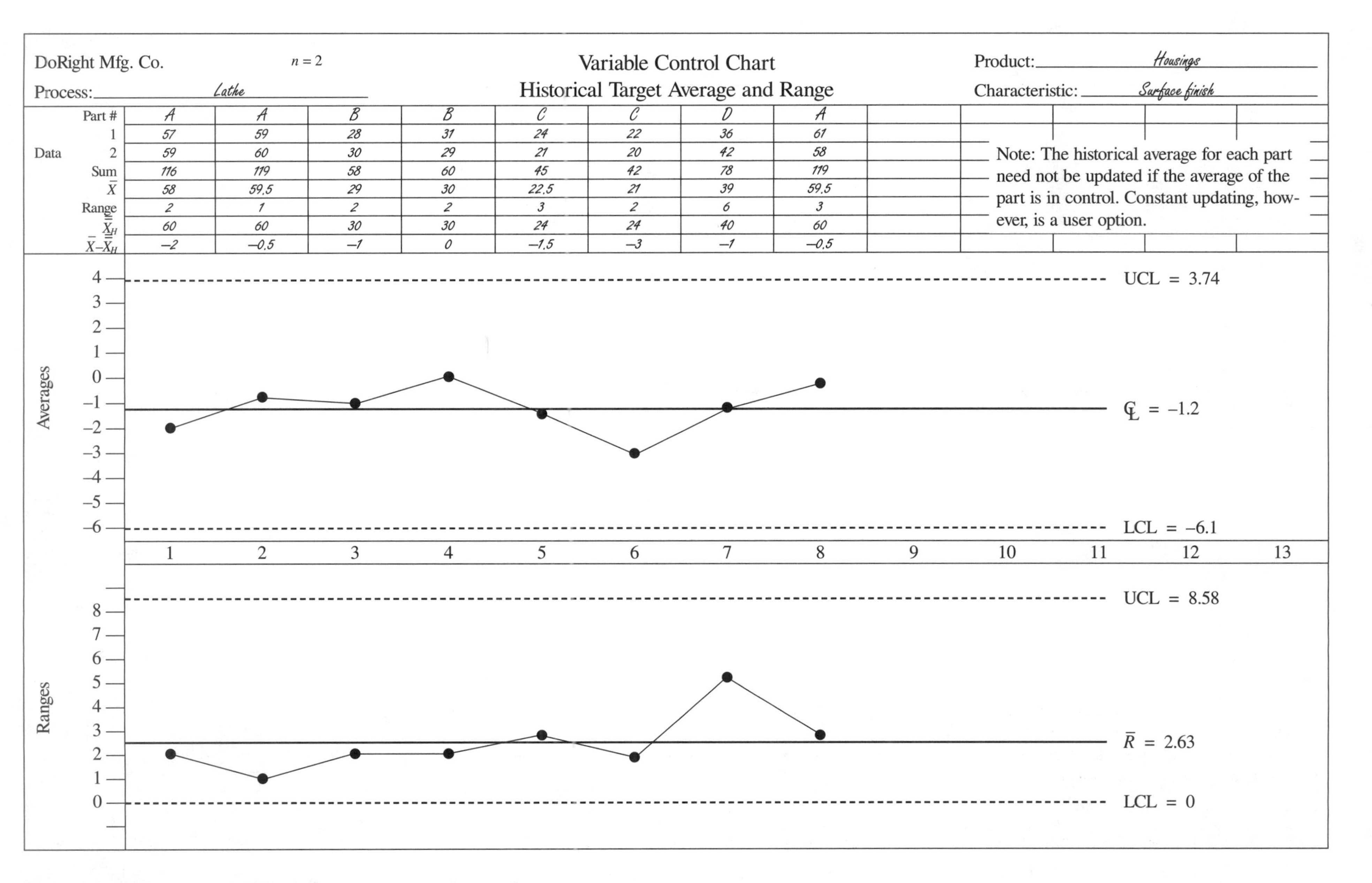

DoRight Mfg. Co. $n = 2$

Process: *Lathe*

Variable Control Chart

Historical Target Average and Range

Product: *Housings*

Characteristic: *Surface finish*

	Part #								
	Part #	A	A	B	B	C	C	D	A
Data	1	57	59	28	31	24	22	36	61
	2	59	60	30	29	21	20	42	58
	Sum	116	119	58	60	45	42	78	119
	$\overline{X}$	58	59.5	29	30	22.5	21	39	59.5
	Range	2	1	2	2	3	2	6	3
	$\overline{\overline{X}}_H$	60	60	30	30	24	24	40	60
	$\overline{X}-\overline{\overline{X}}_H$	−2	−0.5	−1	0	−1.5	−3	−1	−0.5

Note: The historical average for each part need not be updated if the average of the part is in control. Constant updating, however, is a user option.

Figure 3.5. Solution to practice historical target average and range chart.

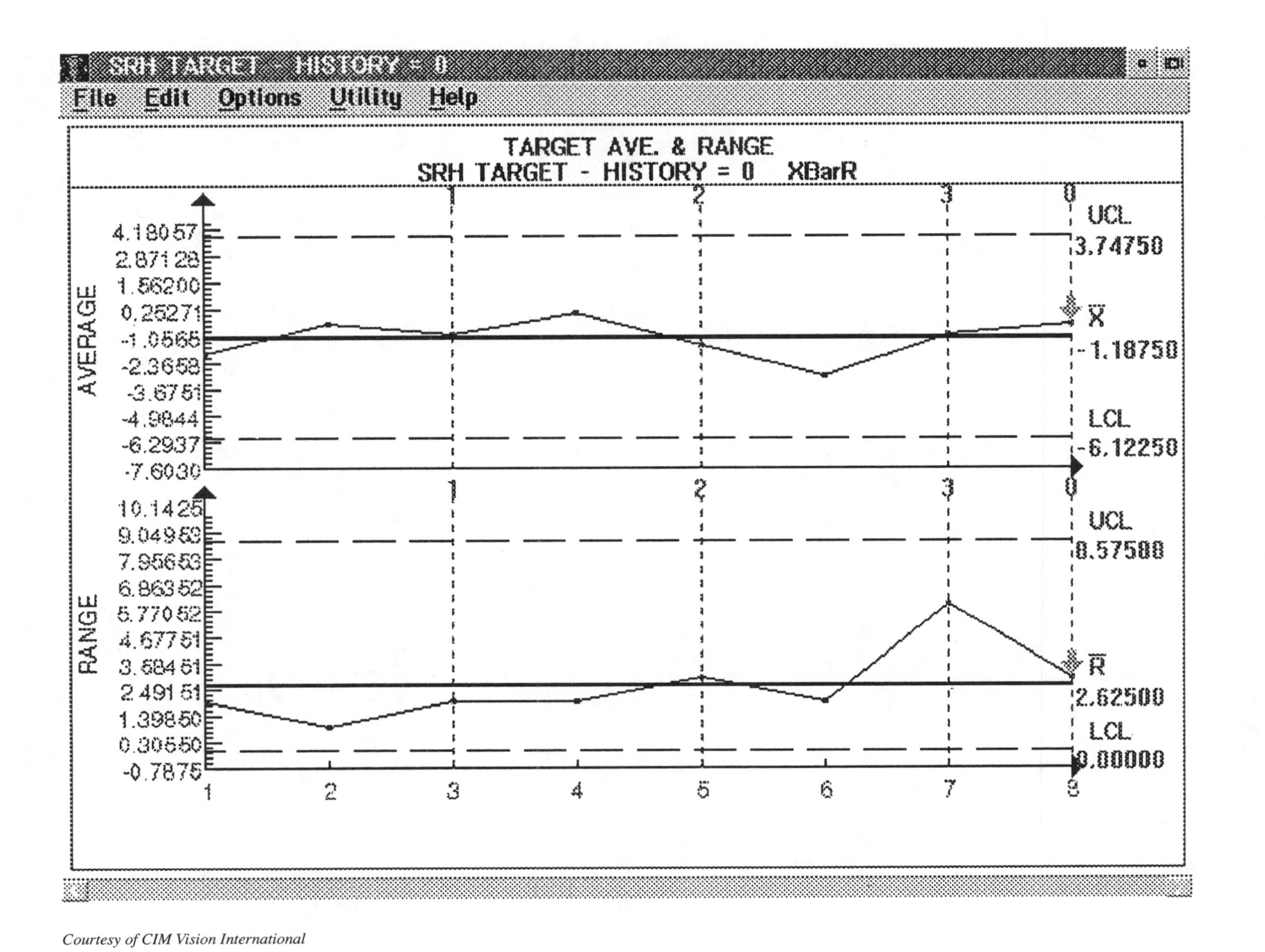

Courtesy of CIM Vision International

Figure 3.6. SQM software example of a historical target average and range chart.

1. Find historical data in order to establish the historical grand average for each different part that will be charted. Record the part numbers and their historical target averages somewhere on the chart for future use. Note: Users have the option to constantly update historical values as new data are obtained. However, these values need not be updated as long as the deviations from historical averages are in control. Also note that variability among different part numbers must be representative (as with the desired target approach), so tests for variability on the ranges still apply.

2. Gather data on each part as it is produced, and record the raw data in the data block of the chart. Also record the part number associated with that data.

3. Compute the sum, average, and range for each subgroup as if it were a traditional average and range chart.

4. For the averages chart, compute the difference between the subgroup average and the historical grand average for each subgroup, and enter the difference in the data block. These are the plot point values for the averages chart. The actual ranges will be plotted on the range chart.

5. Compute the centerline of the averages chart. This particular chart is different than all other averages charts because the centerline is the grand average of the deviations of subgroup averages from their respective historical grand averages. The centerline of the averages chart, in other words, is the average of the subgroup deviations.

$$\text{Centerline} = \frac{\Sigma\,(\bar{x} - \bar{\bar{x}}_H)}{k}$$

where k = the number of subgroups

6. Compute the upper and lower control limits for the averages chart using the following equation.

$$\text{UCL/LCL} = \text{Centerline} \pm A_2 \bullet \bar{R}$$

7. Compute the centerline and control limits for the range chart in the traditional manner.

8. Plot only the deviations from the historical grand average on the averages chart, and plot the ranges on the range chart. Then interpret the charts for control. Note: It is important to highlight part number changes on the chart for analysis and reporting purposes. This is usually accomplished with a vertical dashed or dotted line at the subgroup where the data for the new part number begins.

How To Interpret Historical Target and Range Charts

If process averages go out of control, this is evidence that there has been change in the variability of the part number(s) since historical data was gathered. It is possible in this charting method to see any of the patterns of out-of-control processes such as trends, runs, cycles, and so on. If there are plot points out of control on the averages (deviations) chart, the process is out of control and should be corrected. Also, if corrected, the historical value need not be updated, but if the cause is not corrected, use the latest data to update the historical grand average.

The range chart must be tested for nonrepresentative variability using either the range test or Fishers *F* test.

Variability Must Be Representative

When using target control charts (desired or historical targets), the variability of each part should be representative of the others. Short-run control charts allow the user to reduce the workload by reducing the number of charts required to control the process. However, short-run charts require that the user watch for parts that do not vary representatively. In order to do this, tests must be performed. One of the simplest tests for representative variation is referred to as the range test.

The Range Test

The range test can be used to compare the ranges of any particular part number that is suspected of varying differently than other part numbers plotted on a target control chart. The user should watch for run patterns on the range chart immediately after a new part number has been set up and data have been plotted. The range test is a matter of comparing the average range ($\bar{R}_{\text{all}}$) of all data (including the suspect part) to the average range ($\bar{R}_{\text{suspect}}$) of the suspect part data. The following equations are used for performing the range test on suspect parts.

If suspect part is worse (high run pattern)

$$\frac{\bar{R}_{\text{suspect}}}{\bar{R}_{\text{all}}} \leq 1.33$$

If suspect part is better (low run pattern)

$$\frac{\bar{R}_{\text{all}}}{\bar{R}_{\text{suspect}}} \leq 1.33$$

If either equation exceeds 1.33, the variability of the part is not representative, and it should be charted separately.

The Fisher F *Test*

Another popular test for representative variability, the Fisher *F* test is considerably more involved than the range test (and also more accurate). The purpose of the *F* test is to test the variances between two samples (using the raw data of those samples) to see if they are, or are not, representative. The two samples in question are the data for the suspected part versus the data for other parts that are representative. Refer to Appendices D and E for more information about the *F* test.

STANDARDIZED VARIABLES CONTROL CHARTS

As mentioned earlier, another method for coding data is to standardize the plot points on the chart with respect to their historical value. All variable control charts and attribute control charts can be standardized. The drawbacks to using standardized charts are

1. When more than one process is plotted on the chart, there is a loss of visibility of the continuous variation of each process.

2. Standardized charts require historical statistics for each part/process that is plotted, and those historical values must be constantly updated.

The benefits of using standardized charts are

1. No tests are needed to test for representative variability.
2. One chart can chart the variation from several processes.
3. The control limits are fixed, so they never need to be recomputed as do traditional control charts. The plot points are standardized to the limits.

Standardized chart applications are

1. The variability of all part numbers to be charted on the same chart is not representative.
2. It is desirable to plot more than one process on a single chart, and the process averages are different.

STANDARDIZED AVERAGE AND RANGE CHARTS

A standardized average and range chart can be used to plot different part numbers and different processes on the same chart regardless of their historical variability. The standardized chart is completely robust to these differences in the parts or processes.

The centerline of the standardized averages control chart is always zero. The centerline represents zero difference between the existing plotted average and the historical average value of that part/process. For example, if a plot point on a standardized chart is at zero (the centerline), this means that the latest value for the average of that part does not deviate from the historical average value.

The centerline of the standardized range chart is always one. This indicates that the range of the latest subgroup has a 1:1 ratio to the historical range for that part.

The control limits for the averages chart are fixed and plus or minus A_2. The UCL for the standardized range chart is fixed at D_4 and the LCL is fixed at D_3.

Plot Points

Each plot point on the standardized averages chart must be computed. First, as with all averages charts, the average of the latest subgroup must be computed. Then, that average must be standardized as follows:

$$\bar{X}_{\text{plot}} = \frac{\bar{X}_A - \bar{X}_{AH}}{\bar{R}_{AH}}$$

where

$\bar{X}_A$ = the average of the current subgroup for part A
$\bar{X}_{AH}$ = the historical average of part A
$\bar{R}_{AH}$ = the historical range of part A

For example, for a recent plot point from part A, the following is known. The average of the recent subgroup is .502." The historical average of part A is .501." The historical range for part A is .002." The plot point value for the averages chart is computed as follows:

$$\bar{X}_{\text{plot}} = \frac{.502 - .501}{.002} = \frac{.001}{.002} = +0.5$$

Each plot point on the standardized range chart must be computed. First, as with all range charts, the range of the latest subgroup must be computed. Then, that range must be standardized as follows:

$$R_{plot} = \frac{R_A}{\bar{R}_{AH}}$$

where

R_A = the range of the current subgroup for part A

$\bar{R}_{AH}$ = the historical average range of part A

For example, for a recent plot point from part A, the following is known. The range of the current subgroup for part A is .001." The historical average range for part A is .0008." The plot point value for the range chart is computed as follows:

$$R_{plot} = \frac{.001}{.0008} = 1.25$$

This means that the latest subgroup range for part A is 1.25 times the historical average range for part A.

For practice on standardized average and range charts, refer to Figures 3.7 and 3.8.

As mentioned earlier, standardized variable control charts have many possible applications where different parts (regardless of their variation) or different processes may be plotted on the same chart. This is so because all plot points on these charts are standardized with respect to historical values of both their averages and ranges. The author believes, however, that standardized variables control charts should be applied with caution because they (1) Require more calculations for each plot point; (2) Require frequent updating of historical values for each process; and (3) Most importantly, they can cause the user to become too distant from each individual process. For example, one standardized variables chart can be used to monitor the output of several variables from several different processes (where the plot points at random from those processes are plotted at different sequences). The lack of succession in plot points from any individual process can cause the user to lose sight of out-of-control signals in that process. If users apply these charts to multiple processes, some imaginative method for plotting points (such as symbols or numbering,) may help the user to see individual process performance (such as trends or runs).

A recommended application for standardized variables charts is one process where the central tendency and variability are a product of the tool. In these cases, the focus is on one process and one variable, but that process has different tools, fixtures, dies, mandrels, and so on that are used. These different tools, fixtures, dies, mandrels, and molds vary differently for every part they produce on one process. Using standardized control charts allows the user to plot all of these different sources of variation on one chart (with respect to their historical values) and still allow the user to focus on the variation of one process. Examples of these processes are

1. A press with different dies for the same characteristic (on different parts). One process is used to produce the parts, but each die that is used affects the central tendency and/or variation of the product.

DoRight Mfg. Co.

Process: Shear

Variable Control Chart
Standardized Average and Range

Product: Various cuts

Characteristic: Length

Part #	A	A	B	B	C	C	D	A					
Data 1	7.962	7.973	8.640	8.642	7.980	7.979	8.634	7.981					
Data 2	7.974	7.982	8.643	8.635	7.974	7.983	8.638	7.980					
Avg.													
Range													
Hist. $\bar{X}$	7.979	7.979	8.644	8.644	7.975	7.975	8.637	7.974					
Hist. $\bar{R}$	.008	.008	.006	.006	.009	.009	.005	.010					

Note: The historical average and range for part A have been updated at this plot point.

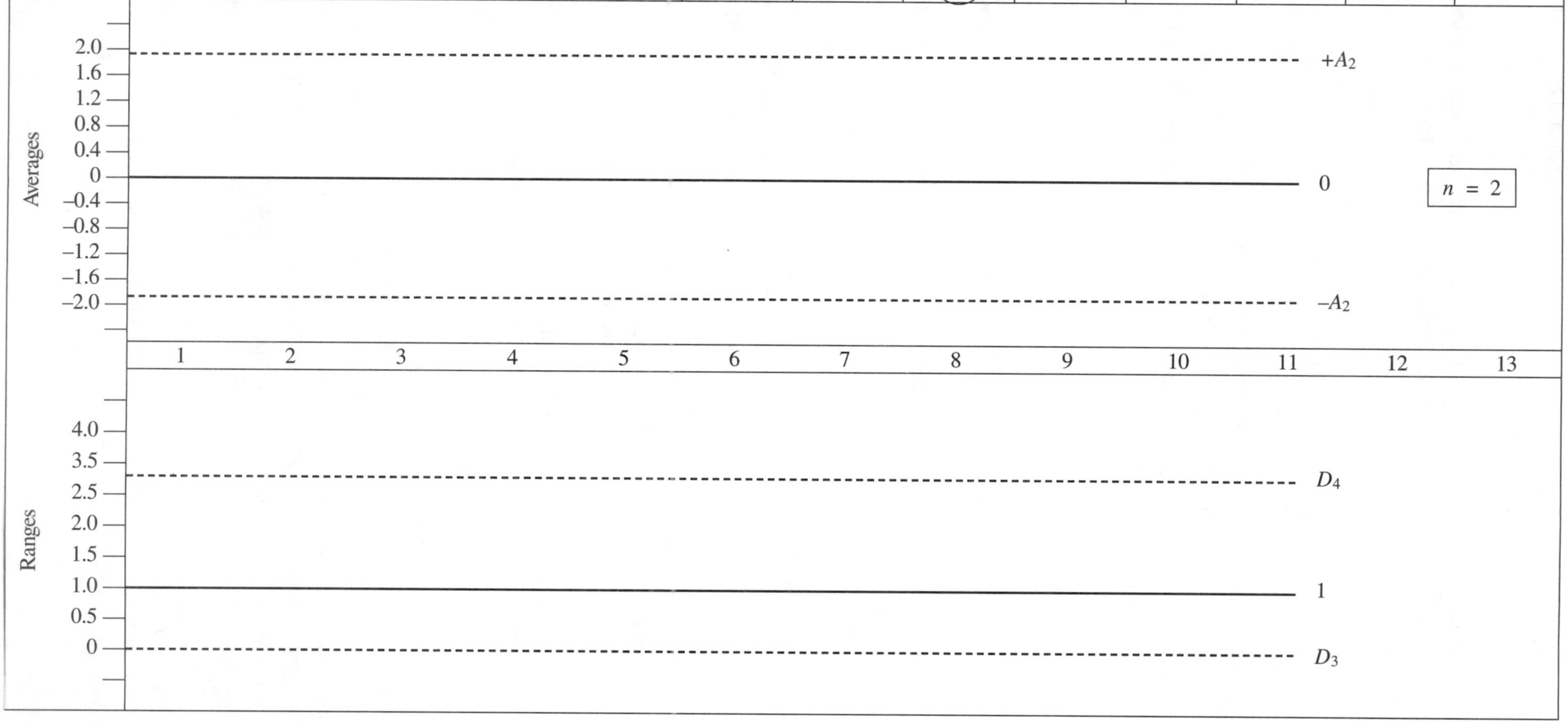

Figure 3.7. Practice standardized average and range chart.

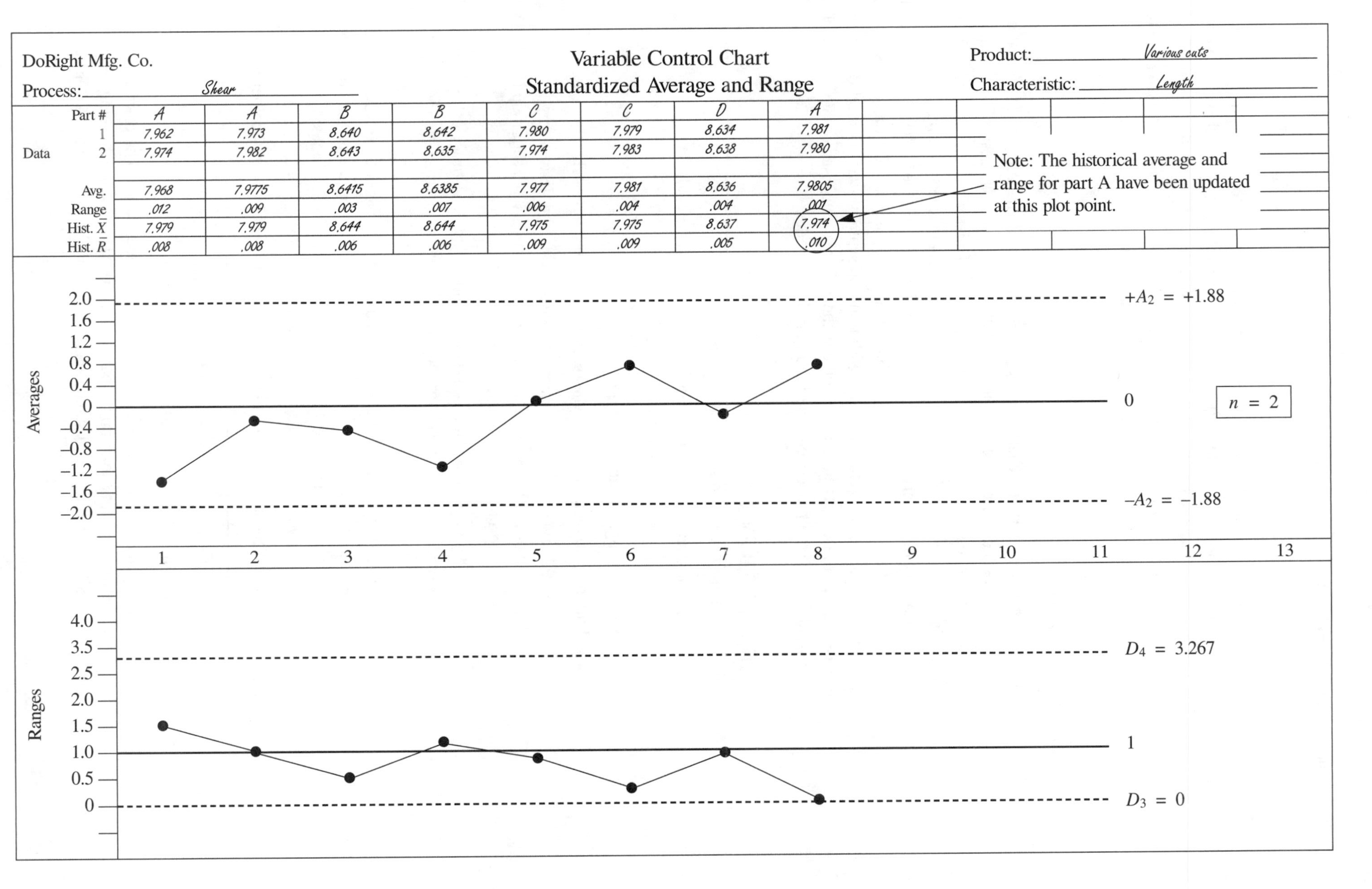

Part #	A	A	B	B	C	C	D	A
Data 1	7.962	7.973	8.640	8.642	7.980	7.979	8.634	7.981
Data 2	7.974	7.982	8.643	8.635	7.974	7.983	8.638	7.980
Avg.	7.968	7.9775	8.6415	8.6385	7.977	7.981	8.636	7.9805
Range	.012	.009	.003	.007	.006	.004	.004	.001
Hist. $\overline{X}$	7.979	7.979	8.644	8.644	7.975	7.975	8.637	7.974
Hist. $\overline{R}$	.008	.008	.006	.006	.009	.009	.005	.010

Figure 3.8. Solution to practice standardized average and range chart.

2. An extrusion machine with different mandrels or dies for each different part. One process is used to produce the parts, but each mandrel or die that is used on the process affects the central tendency and/or variation of the product.

In these cases, historical values for the averages and ranges from these dies or mandrels can be standardized and plotted on one chart, and the user is still focusing on the process for process control.

SUMMARY

The most effective short-run charts are the target charts, and virtually any of the traditional variables control charts can be coded and used in target mode. Use desired target charts for adjustable processes where desired targets exist, and historical target charts for non-adjustable processes, or where there is no desired target. Keep in mind that, on target charts, variability of all part numbers combined (or pooled) on these charts must be representative. The standardized method of coding data is very useful when one chooses to plot different processes on one chart or parts with nonrepresentative variation on one chart. In all cases, short-run charts help us to achieve the primary purposes of SPC: control the process, and make the process capable for all products produced.

REVIEW PROBLEMS

Refer to Appendix H for answers.

1. Find the average of the following coded data.

   9 7 –3 0 –2 3

2. Find the range of the data in question 1.
3. Find the sample standard deviation of the data in question 1.
4. A desired target chart for averages is being used. The desired target is a nominal dimension of .500. The average of three measurements in a subgroup is .497. What is the appropriate plot point value?
5. A historical target chart for averages is being used. The historical average of the part is .752. The actual data for the subgroup has an average of .754. What is the appropriate plot point value?
6. A short-run target chart is being used and a high-run pattern (above the centerline) appears on the range chart immediately after setting up a new part number. The average range of the suspect part is .002, and the overall average range is .004. Is the variability of the new part representative?
7. One of the most important rules about short-run SPC is to remember to focus on the __________________________________.
8. What is the centerline for all standardized range charts?
9. What is the UCL for a standardized averages chart?
10. A short-run standardized chart for averages is being used. The calculated average of the most recent subgroup for part A is .750.″ The historical average for part #A is .748″ and the historical standard deviation for part A is .0008.″ What is the plot point value?

CHAPTER 4

Traditional Attributes Control Charts

INTRODUCTION

Attribute characteristics are characteristics of a product that are evaluated in terms of whether they are good or bad, on or off, go or no-go, and so on. Attribute charts apply when they must be used (a strict attribute, such as the part is cracked or not cracked) or when you choose to use them (the part is measurable, but you choose to gage the part with a go/no-go gage). There are various kinds of defects such as porosity, cracks, surface flaws, and missing parts, depending on the product and its particular characteristics. Note, at this point, the term *nonconformity* is synonymous with defect, and the term *nonconformance* is synonymous with defective. A nonconformity (defect) is a single characteristic of the part. Each part can have several nonconformities. A nonconformance (defective) is a part that has one or more nonconformities. Attribute charts can be used on any attribute that is counted, or one that represents a fraction or percentage.

An important consideration before starting any attribute chart is to make sure that the purpose of the chart is defined and that the standards are available and understood. Various kinds of attribute control charts may be used. The chart that is selected depends on the specific attributes to be controlled. The attribute charts covered in this book include the following:

p chart—controls the fraction nonconforming (defective)

np chart—controls the number nonconforming (defective)

c chart—controls the number of nonconformities (defects)

u chart—controls the number of nonconformities (defects) per unit

The equations for the control limits of these attribute charts are shown in Figure 4.1.

BREAKTHROUGH IMPROVEMENTS WITH ATTRIBUTE CHARTS

Attribute charts, by themselves, only provide the users a monitoring device. They help users to identify the chronic level of poor performance of the process (the process average) and whether that level of poor performance is stable or not (in control or out of control). For example, if a *p* chart shows that the process is in statistical control with a process average ($\bar{p}$) of .10, this means that the process is consistently producing .10 proportion (or 10 percent) defective products.

$$\bar{p} \pm 3\sqrt{\frac{\bar{p}\,(1 - \bar{p})}{n}} \qquad \bar{c} \pm 3\sqrt{\bar{c}}$$

$$\bar{u} \pm 3\,\frac{\sqrt{\bar{u}}}{\sqrt{n}} \qquad n\bar{p} \pm 3\sqrt{n\bar{p}\left(1 - \frac{n\bar{p}}{n}\right)}$$

Figure 4.1. Attribute chart equations.

Attribute control charts are not as sensitive to variation as variable charts. Control over a process is important, but there are ways to increase the effectiveness of attribute charts where breakthrough improvements are made. *Breakthrough improvement* refers to any improvement in a process that significantly reduces the process average (or chronic level of poor performance) of that process. In this case, *chronic problems,* or recurring problems, are often a large proportion of the poor performance of a process. *Sporadic problems,* on the other hand, are isolated problems (one-time occurrence) that, when solved, have a very small impact on reducing the level of poor performance of the process.

Any attribute chart can be a very powerful tool when accompanied by Pareto analysis (see chapter 10). Pareto analysis helps to identify the most significant contributors to the process average ($\bar{p}$, $\bar{c}$, $n\bar{p}$, $\bar{u}$) and the centerline values of p, c, np, and u charts, respectively. When one of the vital few contributors are solved (via problem-solving teams and methods), the process average of poor performance drops significantly.

Breakthrough Example

A shaft production process is in control (using a p chart) and the process average is .12 (or 12 percent defective output). In conjunction with the p chart, a Pareto analysis is also plotted on all of the individual defect types involved with the shaft production process. In this manner, the process average ($\bar{p}$ = .12) is fully explained by the defects on the Pareto chart. There are six defects involved: oversize diameter, undersize diameter, bent shafts, cracked shafts, runout, and surface finish. Table 4.1 shows the results of the Pareto chart.

Keep in mind that the total number of defects (2851) is directly correlated to the process average (.12 or 12 percent defective). The top contributor, as shown in Table 4.1, is bent shafts at 42.2 percent of total defects. The smallest contributor, cracked shafts, represents only 3.9 percent of total defects. If the top problem, bent shafts, were solved by a team of personnel using problem-solving methods (see chapter 10), the process average (12 percent) would go down to 7 percent immediately. If the team works on the smallest contributor, cracked shafts (representing only 3.9 percent of the 12 percent defective average), the improvement would be quite small (down to 11.5 percent). Solving the problem of bent shafts would cause a breakthrough improvement, while solving the problem of cracked shafts would have very little effect on improving the shaft line performance.

Table 4.1. Pareto analysis of shaft line defects.

Defect	Total quantity rejected	Percent of total defects
Oversize diameter	165	5.8 %
Undersize diameter	137	4.8%
Bent shafts	1203	42.2%
Cracked shafts	112	3.9%
Poor runout	1110	38.9%
Poor finish	124	4.3%

The solution of problems in any company has a lot to do with the resources available for problem solving. When there are few resources (people, capital, and so on), the company needs to focus on breakthrough improvements. Pareto analysis along with the control chart can focus attention on the problems that get the most return for the effort. If, for example, the company can assemble two problem-solving teams, the logical thing to do is work on the top two contributors. In this example, they are bent shafts (42.2 percent) and poor runout (38.9 percent). Together, these two problems constitute 81.1 percent of total defects (also 81.1 percent of 12 percent defective). If each team concentrated on the solution of one problem and both teams solved the two problems, the process average (12 percent) would immediately go down to a breakthrough level of 2.3 percent.

MULTIPLE PARETO CHARTS

I have a basic rule when it comes to Pareto analysis. The rule is "never do only one Pareto analysis." The meaning behind this rule is based on the fact that one Pareto analysis can be misleading. For example, a Pareto analysis on a process may reflect significant contributor defects, but does not consider the cost of those defects. Hence, teams are working on improving a defect on a part that is worth $5, when another defect on another part (worth $300) is ignored. Using just one Pareto analysis, in this case, leads the team to solve a problem that really doesn't cost too much and ignore a problem that is losing the company a lot of money.

This isn't to say that teams should always focus on high-cost items or on product functional items. The point is that using more than one Pareto analysis gives the company and the teams all the information required to make sound decisions on the problems that need to be solved.

THE *p* CHART—FRACTION NONCONFORMING (DEFECTIVE)

The most commonly used attribute control chart in industry is the p chart. The p chart is used specifically to control the fraction or proportion nonconforming. The *fraction nonconforming* is the ratio of the quantity nonconforming divided by the quantity of parts inspected.

For example, where

Quantity of nonconforming parts = 3

Number of parts inspected = 20

$$p = \frac{3}{20} = 0.15$$

The p chart is used in the shop for several reasons, some of which are as follows:

1. On processes that use go/no-go gaging such as plug gages, ring gages, and functional gages
2. In inspection areas to monitor or control the fraction nonconforming from the shop or from suppliers
3. In situations where visual inspection is performed
4. Other attribute situations where the fraction (or proportion) nonconforming requires control

The p chart can also be used in office settings for

1. Nonconforming paperwork
2. Fraction (or the percent) completed jobs
3. Fraction (or percent) time spent

Steps to Construct a p *Chart*

Refer to Figures 4.2, 4.3, and 4.4 for practice.

1. Decide the specific purpose of the chart.
2. Decide what areas you want to control.
3. Choose the sample size. The sample size should be constant, if possible. If sample sizes vary slightly, the average sample size ($\overline{n}$) can be used to calculate control limits. If sample sizes vary more than 25 percent from the average sample size, control limits must be calculated around each plotted point. See the steps to follow.
4. Record the sample size (n) and the number of nonconformances found (np).
5. Calculate p for each sample.

$$p = \frac{np}{n} = \frac{\text{Number of nonconformances found}}{\text{Number of parts inspected}}$$

6. Find the average fraction nonconforming ($\overline{p}$), which is the centerline of the control chart. Do not add the p values and divide unless sample sizes are equal or constant.

$$\overline{p} = \frac{\text{The sum of all nonconformances found}}{\text{The sum of all samples taken}}$$

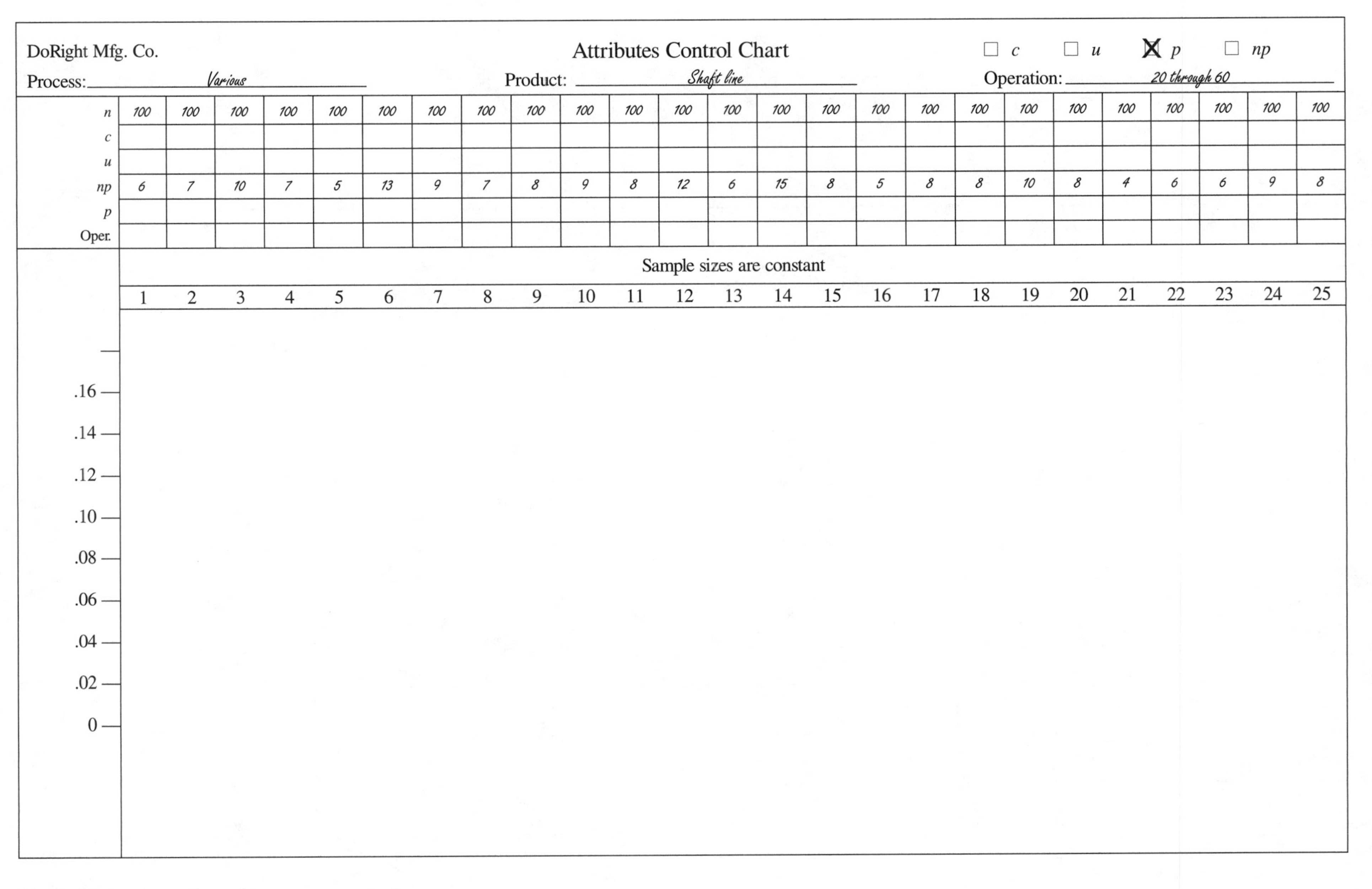

DoRight Mfg. Co.

Attributes Control Chart

☐ c ☐ u ☒ p ☐ np

Process: Various

Product: Shaft line

Operation: 20 through 60

	1	2	3	4	5	6	7	8	9	10	11	12	13	14	15	16	17	18	19	20	21	22	23	24	25
n	100	100	100	100	100	100	100	100	100	100	100	100	100	100	100	100	100	100	100	100	100	100	100	100	100
c																									
u																									
np	6	7	10	7	5	13	9	7	8	9	8	12	6	15	8	5	8	8	10	8	4	6	6	9	8
p																									
Oper.																									

Sample sizes are constant

Figure 4.2. Practice p chart with constant sample sizes.

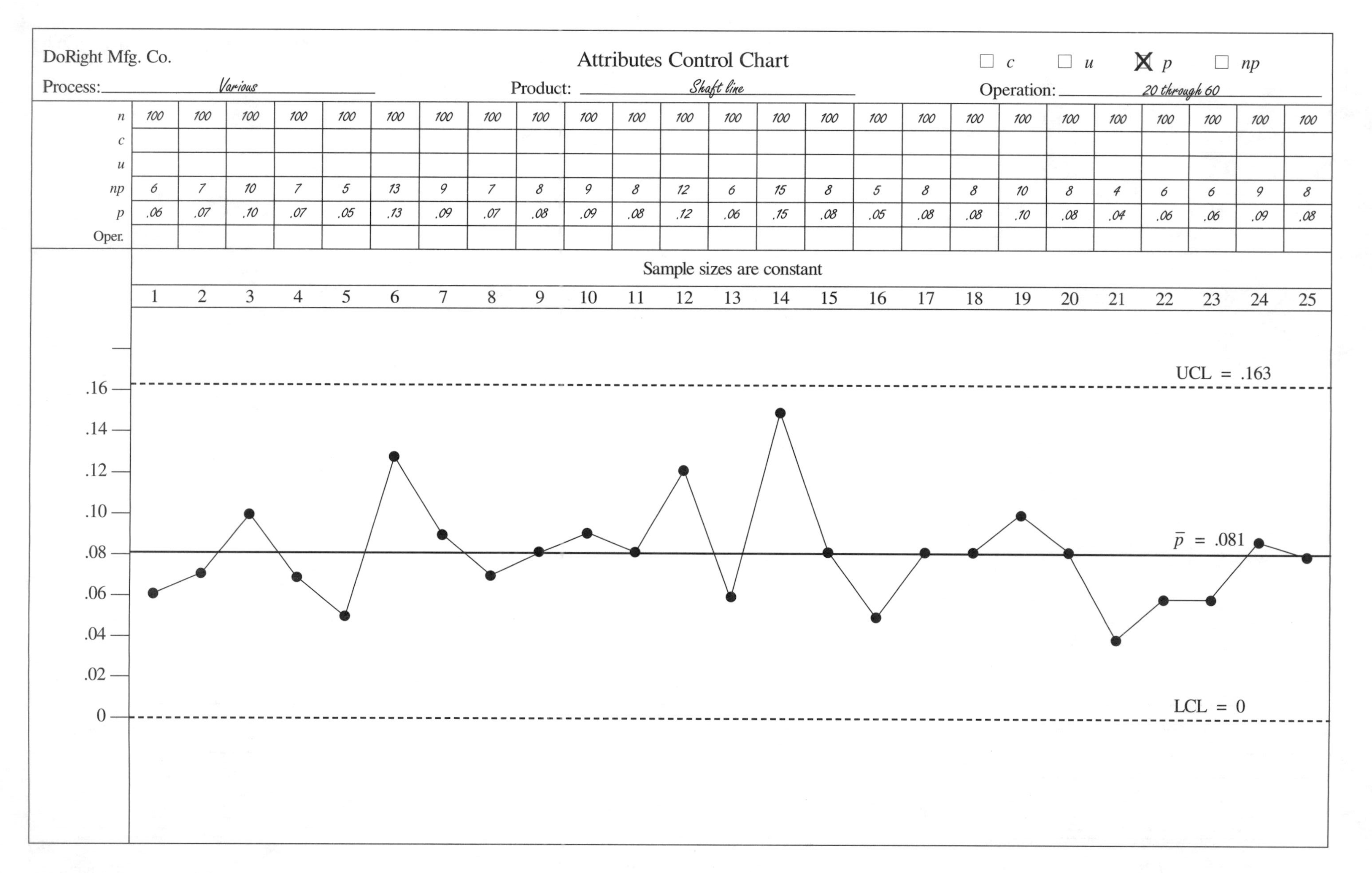

	1	2	3	4	5	6	7	8	9	10	11	12	13	14	15	16	17	18	19	20	21	22	23	24	25
n	100	100	100	100	100	100	100	100	100	100	100	100	100	100	100	100	100	100	100	100	100	100	100	100	100
c																									
u																									
np	6	7	10	7	5	13	9	7	8	9	8	12	6	15	8	5	8	8	10	8	4	6	6	9	8
p	.06	.07	.10	.07	.05	.13	.09	.07	.08	.09	.08	.12	.06	.15	.08	.05	.08	.08	.10	.08	.04	.06	.06	.09	.08
Oper.																									

Figure 4.3. Solution to practice *p* chart with constant sample sizes.

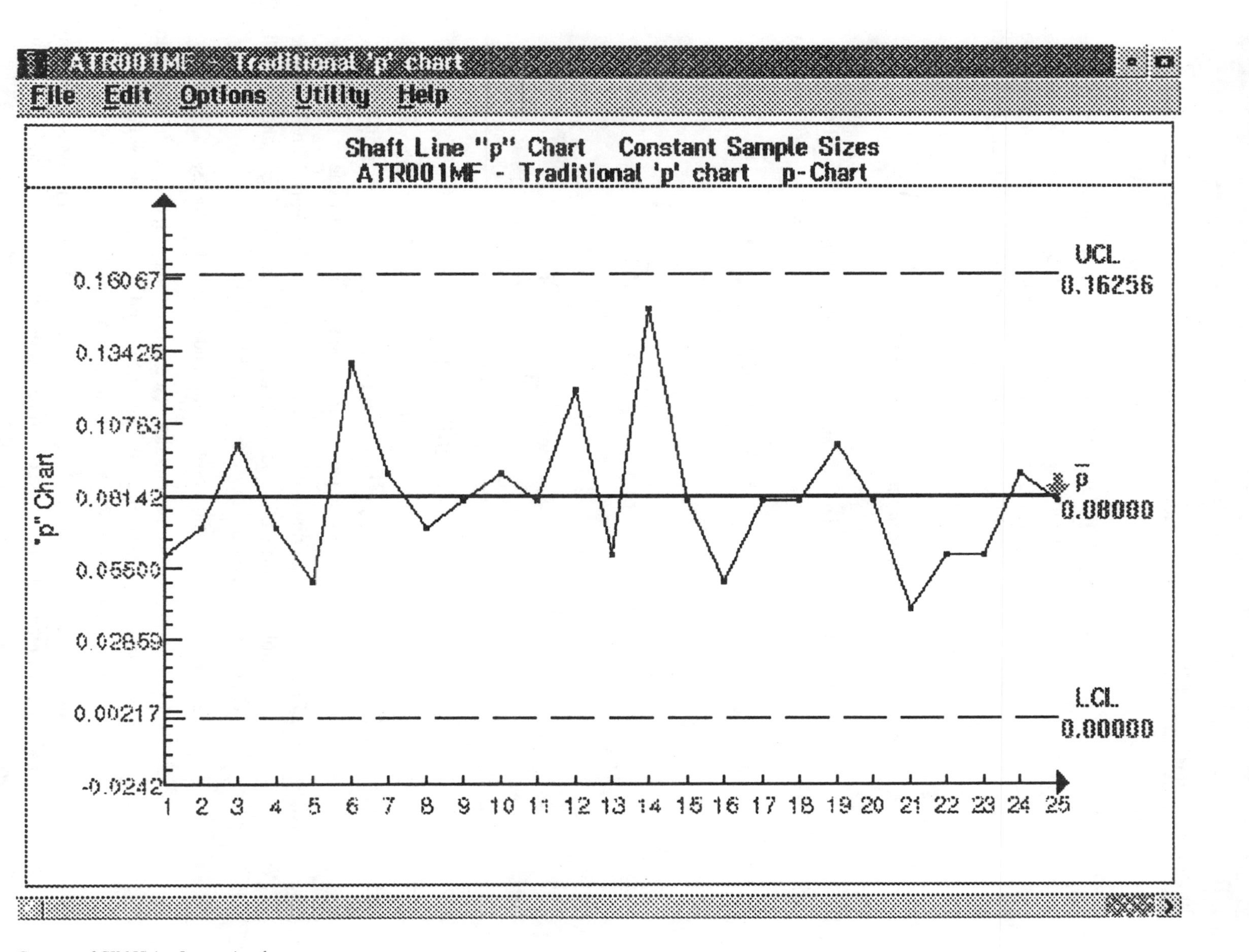

Courtesy of CIM Vision International

Figure 4.4. SQM software example of a p chart with constant sample sizes.

7. Calculate the UCL and the LCL for the p chart. If samples (n) do not vary, use n in the formula. If samples (n) vary less than 25 percent from the average sample size, use ($\bar{n}$) in the formula. If samples vary over 25 percent from the average sample size, control limits must be calculated around each plotted point on the chart.

$$\text{Control limits} = \bar{p} \pm 3\left(\sqrt{\frac{\bar{p}\,(1-\bar{p})}{n}}\right)$$

Example The sample sizes are constant (n = 100) and p is 0.26.

$$CL = \bar{p} \pm 3\left(\sqrt{\frac{\bar{p}\,(1-\bar{p})}{n}}\right)$$

$$= .26 \pm 3\left(\sqrt{\frac{.26\,(1-.26)}{100}}\right)$$

$$= .26 \pm 3\left(\sqrt{\frac{.1924}{100}}\right)$$

$$= .26 \pm 3\,(.044)$$

$$= .26 \pm .132$$

$$UCL = .26 + .132 = .392 \qquad LCL = .26 - .132 = .128$$

Note: In some cases, the LCL will be a negative answer. If so, the lower control limit is zero. This is true particularly when $\bar{p}$ is a small value.

8. Plot ($\bar{p}$) the centerline and the control limits on the chart. The $\bar{p}$ goes near the center of the chart (solid line) and the scale is selected using $\bar{p}$ and the UCL. The control limits are usually plotted in dashed lines.
9. Plot each p value (point) on the chart and connect the points.
10. Look for out-of-control conditions (see pattern analysis, chapter 6). If you find and eliminate them, recalculate the centerline and control limits excluding all of the data for the points which were out of control.

When a p chart is in control, the process average ($\bar{p}$) is used to assess capability. The previous chart reflected an average of .26 (or 26 percent defective); therefore, the yield of good parts that can be expected from this process is only 74 percent.

How to Interpret the p *Chart*

A point above the UCL means

1. The process has worsened either at that point in time or as part of a trend.
2. A mistake may have been made in control limit calculations or plotting the point.

3. The measuring system (the gage, method, or observer) may have changed.
4. The standards may have changed or may not be clear.

A point below the LCL means

1. The process has improved. Note: You should still find the special cause for this. You want to keep doing things that cause improvements.
2. The control limit or plotted point is in error.
3. The measuring system (gage, method, observer) has changed.

Control Limits for Varying Sample Sizes

Refer to practice chart in Figures 4.5, 4.6, and 4.7 for varying samples sizes.

1. Each plotted point (p) and the centerline ($\bar{p}$) are the same as discussed previously.
2. Calculate 3 sigma$_p$ using $\bar{p}$ and remember it. The 3 sigma$_p$ value doesn't change unless $\bar{p}$ changes.

$$3\sigma_p \pm 3\left(\sqrt{\bar{p}\ (1 - \bar{p})}\right)$$

3. The control limits around each plotted point are calculated as follows:

$$\bar{p} \pm \frac{3\sigma_p}{\sqrt{n}}$$

where n is the sample size for the subgroup.

THE *np* CHART—NUMBER OF NONCONFORMANCES (DEFECTIVES)

The np chart is used to monitor the number of nonconformances (np) rather than the fraction nonconforming (p). As with the p chart, sample sizes should be constant if possible. The np chart is usually used when the number of nonconformances is easier and more meaningful to report; and the sample size remains constant from period to period.

Steps to Construct the np *Chart*

Refer to Figures 4.8, 4.9, and 4.10 for practice.

1. Record the constant sample sizes on the chart.
2. Record the number of nonconformances found in each sample.
3. Plot the number of nonconformances for each sample on the chart.
4. Calculate the average number of nonconformances ($n\bar{p}$).

$$n\bar{p} = \frac{\text{The sum of all nonconformances}}{\text{The total number of subgroups}}$$

5. Calculate the upper and lower control limits (UCL + LCL).

$$n\bar{p} \pm 3\sqrt{n\bar{p}\left(1 - \frac{n\bar{p}}{n}\right)}$$

DoRight Mfg. Co.

Process: Various

Attributes Control Chart

Product: Housing line

☐ c ☐ u ☒ p ☐ np

Operation: Final inspection

	1	2	3	4	5	6	7	8	9	10	11	12	13	14	15	16	17	18	19	20	21	22	23	24	25
n	60	41	49	110	80	70	66	43	58	21	102	85	62	60	50	50	43	86	60	62					
c																									
u																						$\bar{p}$ =	.029		
np	1	1	2	7	2	3	4	1	1	2	1	2	1	1	1	2	1	1	2	1		σ_p =	.503		
p																									
Oper.																									

.16

.14

.12

.10

.08

.06

.04

.02

0

Figure 4.5. Practice p chart with varying sample sizes.

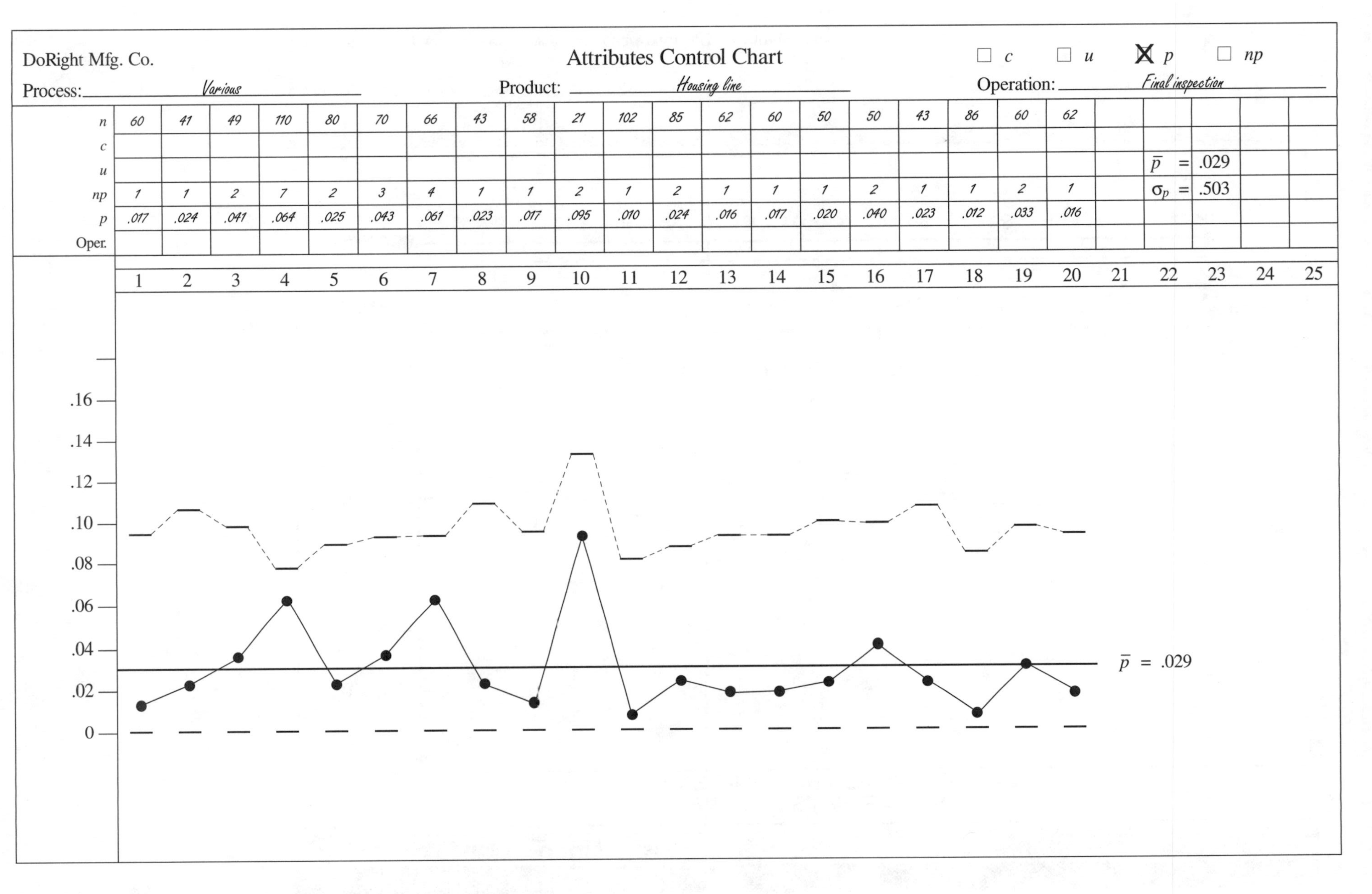

Figure 4.6. Solution to practice p chart with varying sample sizes.

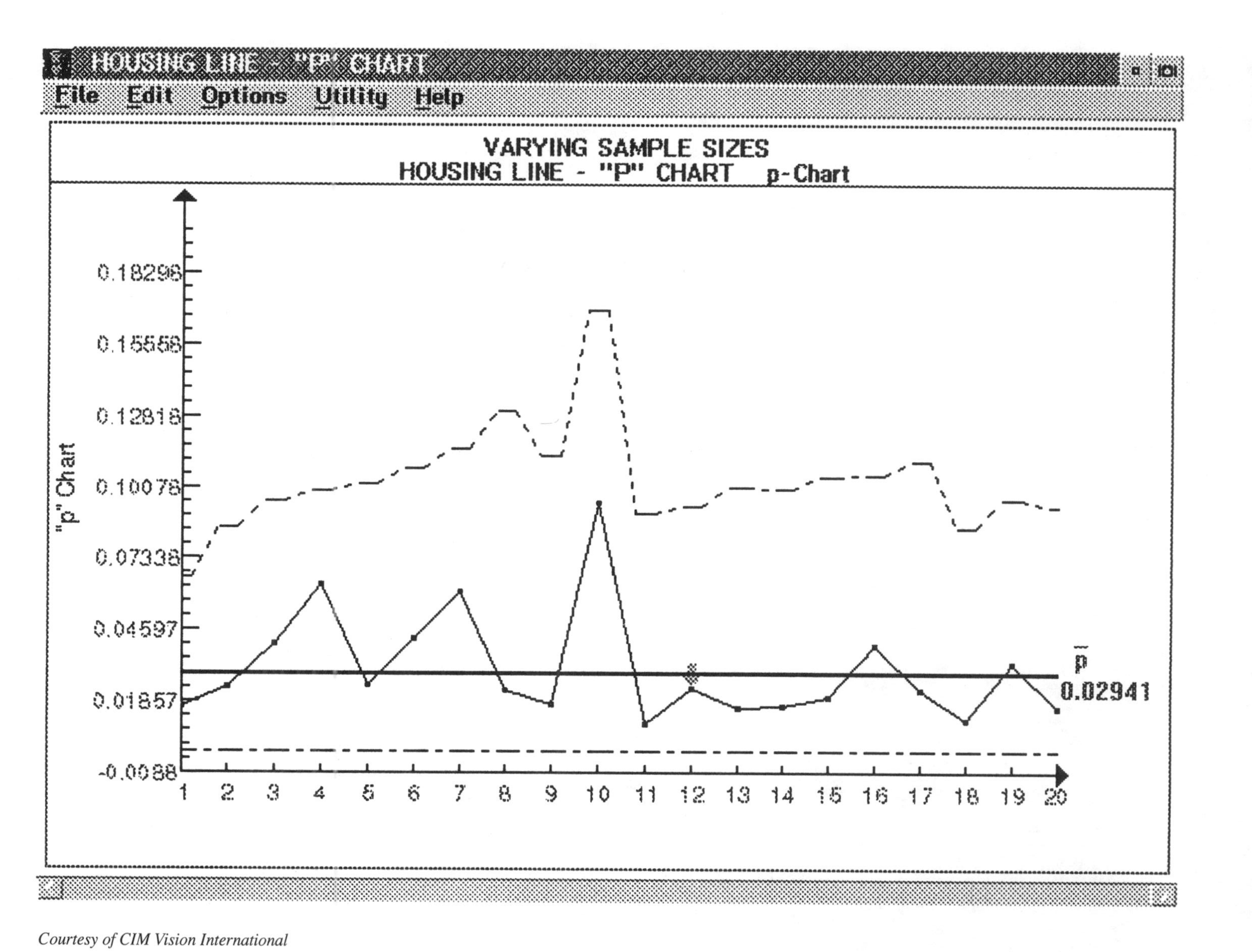

Courtesy of CIM Vision International

Figure 4.7. SQM software example of a p chart with varying sample sizes.

DoRight Mfg. Co.

Attributes Control Chart

☐ c ☐ u ☐ p ☒ np

Process: Shear and press Product: Housing clamp Operation: Final inspection

	1	2	3	4	5	6	7	8	9	10	11	12	13	14	15	16	17	18	19	20	21	22	23	24	25
n	30	30	30	30	30	30	30	30	30	30	30	30	30	30	30	30	30	30	30	30					
c																									
u																									
np	1	3	2	1	2	3	2	0	4	2	2	1	1	2	3	2	5	2	3	2					
p																									
Oper.																									

8
7
6
5
4
3
2
1
0

Figure 4.8. Practice np chart.

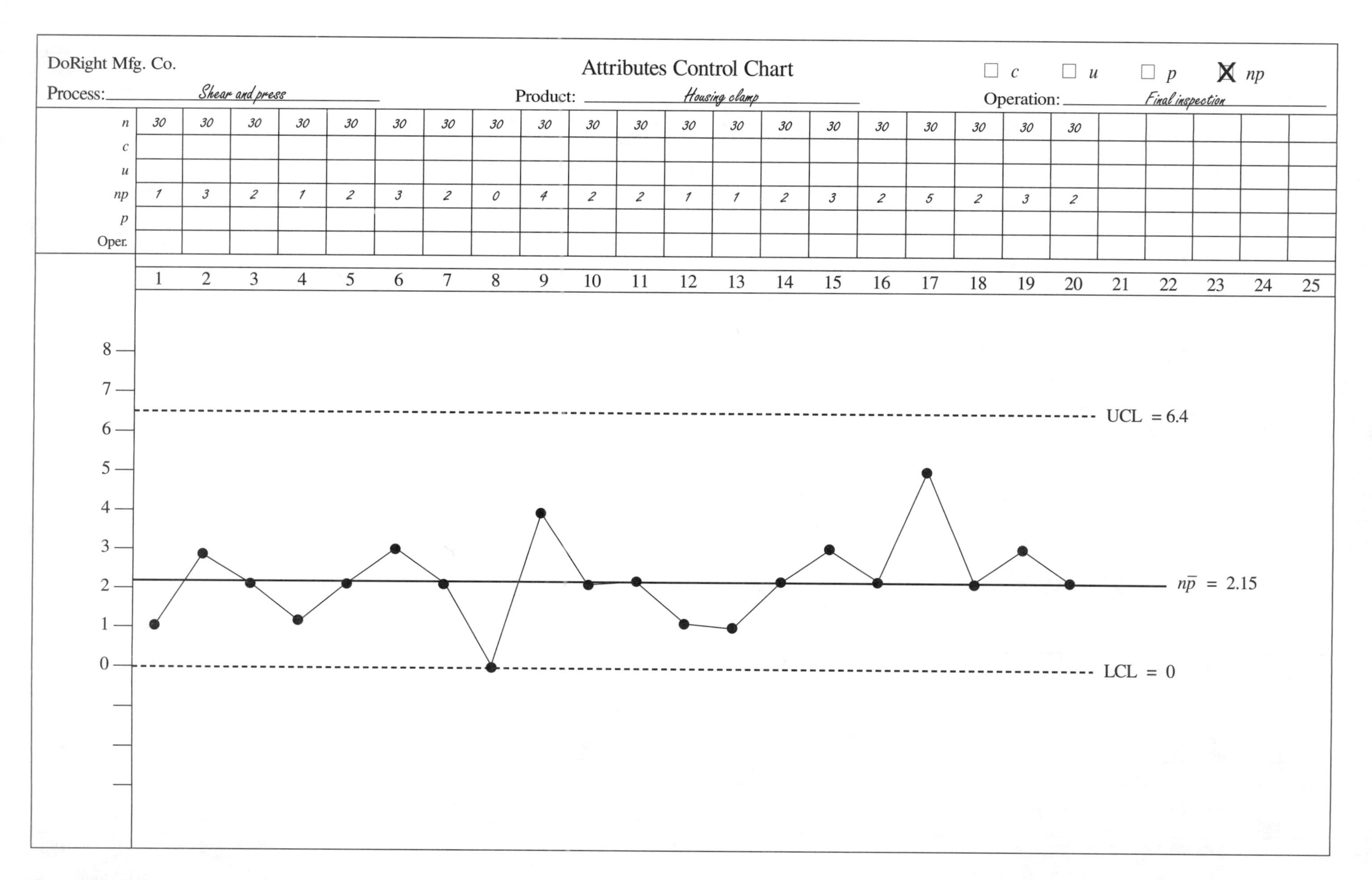

	1	2	3	4	5	6	7	8	9	10	11	12	13	14	15	16	17	18	19	20	21	22	23	24	25
n	30	30	30	30	30	30	30	30	30	30	30	30	30	30	30	30	30	30	30	30					
c																									
u																									
np	1	3	2	1	2	3	2	0	4	2	2	1	1	2	3	2	5	2	3	2					
p																									
Oper.																									

Figure 4.9. Solution to practice *np* chart.

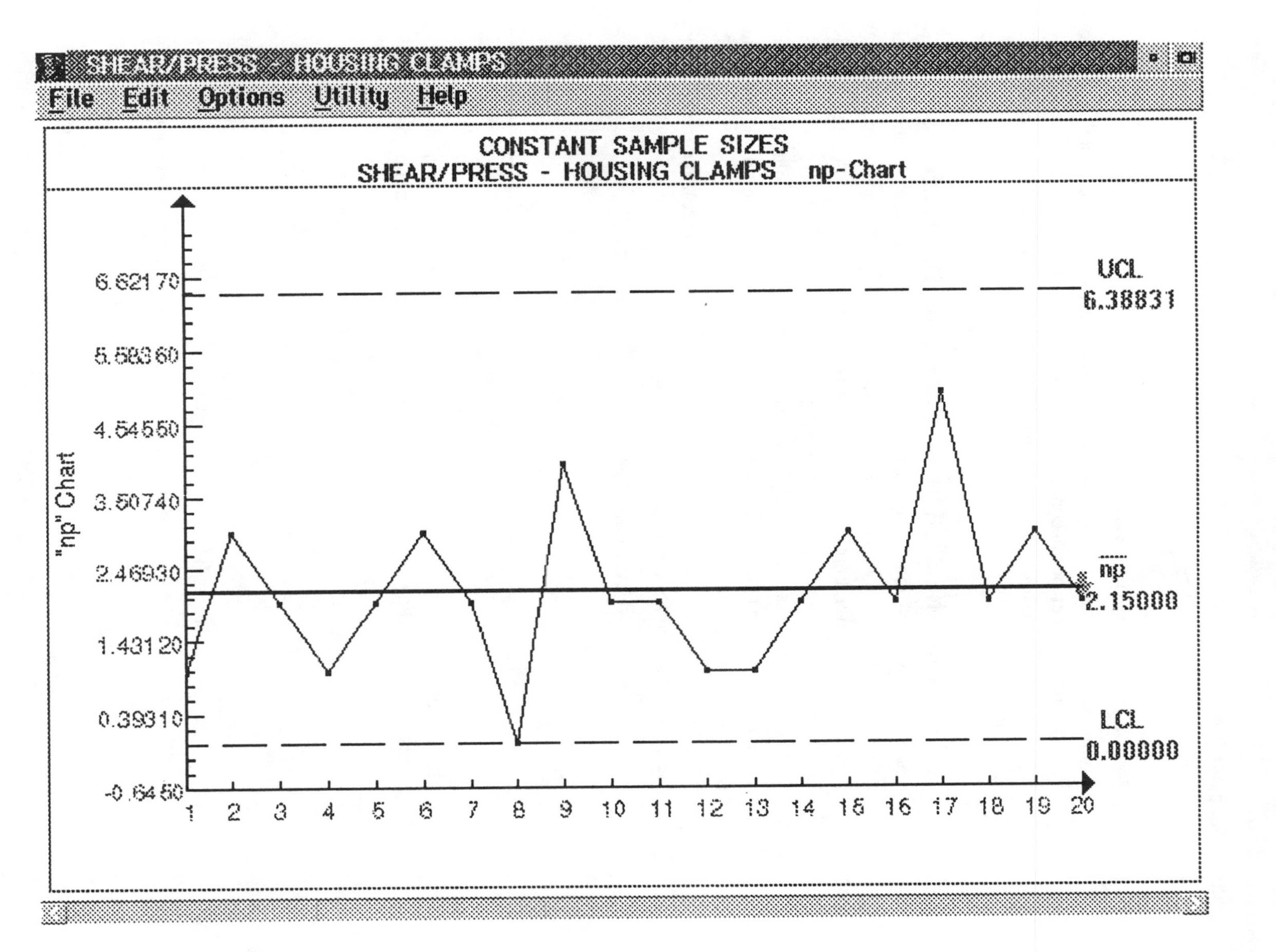

Courtesy of CIM Vision International

Figure 4.10. SQM software example of a *np* chart.

How to Interpret the np *Chart*

Control in the *np* chart is interpreted the same way as the *p* chart except that you are working with the number of nonconformances, not the fraction nonconforming. The process capability is the average number of nonconformances ($n\bar{p}$) when the process is in statistical control.

THE *c* CHART—NUMBER OF NONCONFORMITIES (DEFECTS)

The *c* chart is an attribute chart that is used to control the number of nonconformities. A *nonconformity* is a single quality characteristic that does not conform to requirements. One part could have several nonconformities, but is still only one nonconforming part. Refer to Figures 4.11, 4.12, and 4.13 for practice.

The centerline of the *c* chart is the average number of nonconformities ($\bar{c}$). This is found by adding the total number of nonconformities found in each sample and dividing by the total number of samples taken.

For example, say five automobiles are inspected (one at a time), and the following number of nonconformities are recorded.

Auto number	Number of defects
1	12
2	7
3	10
4	3
5	1

Then, *c* for each automobile is 12, 7, 10, 3, and 1. The average number of nonconformities ($\bar{c}$) is found by adding all the nonconformities and dividing the sum by 5.

$$\bar{c} = \frac{12 + 7 + 10 + 3 + 1}{5}$$

$$= \frac{33}{5}$$

$\bar{c}$ = 6.6 (This is the centerline of the control chart.)

The control limits for the *c* chart are calculated using the following formulas.

$$\begin{aligned} \text{UCL} &= \bar{c} + 3\sqrt{\bar{c}} \\ &= 6.6 + 3\sqrt{6.6} \\ &= 6.6 + 3(2.569) \\ &= 6.6 + 7.707 \\ &= 14.307 \end{aligned} \qquad \begin{aligned} \text{LCL} &= \bar{c} - 3\sqrt{\bar{c}} \\ &= 6.6 - 3\sqrt{6.6} \\ &= 6.6 - 3(2.569) \\ &= 6.6 - 7.707 \\ &= 0 \end{aligned}$$

Note: The LCL is zero because the solution is –1.107 (which is replaced with zero).

After the centerline and control limits are calculated and plotted, the individual *c* values are plotted and the points connected.

When out-of-control conditions are found, there should be an investigation for the particular nonconformities that caused the out-of-control condition. It is likely that specific

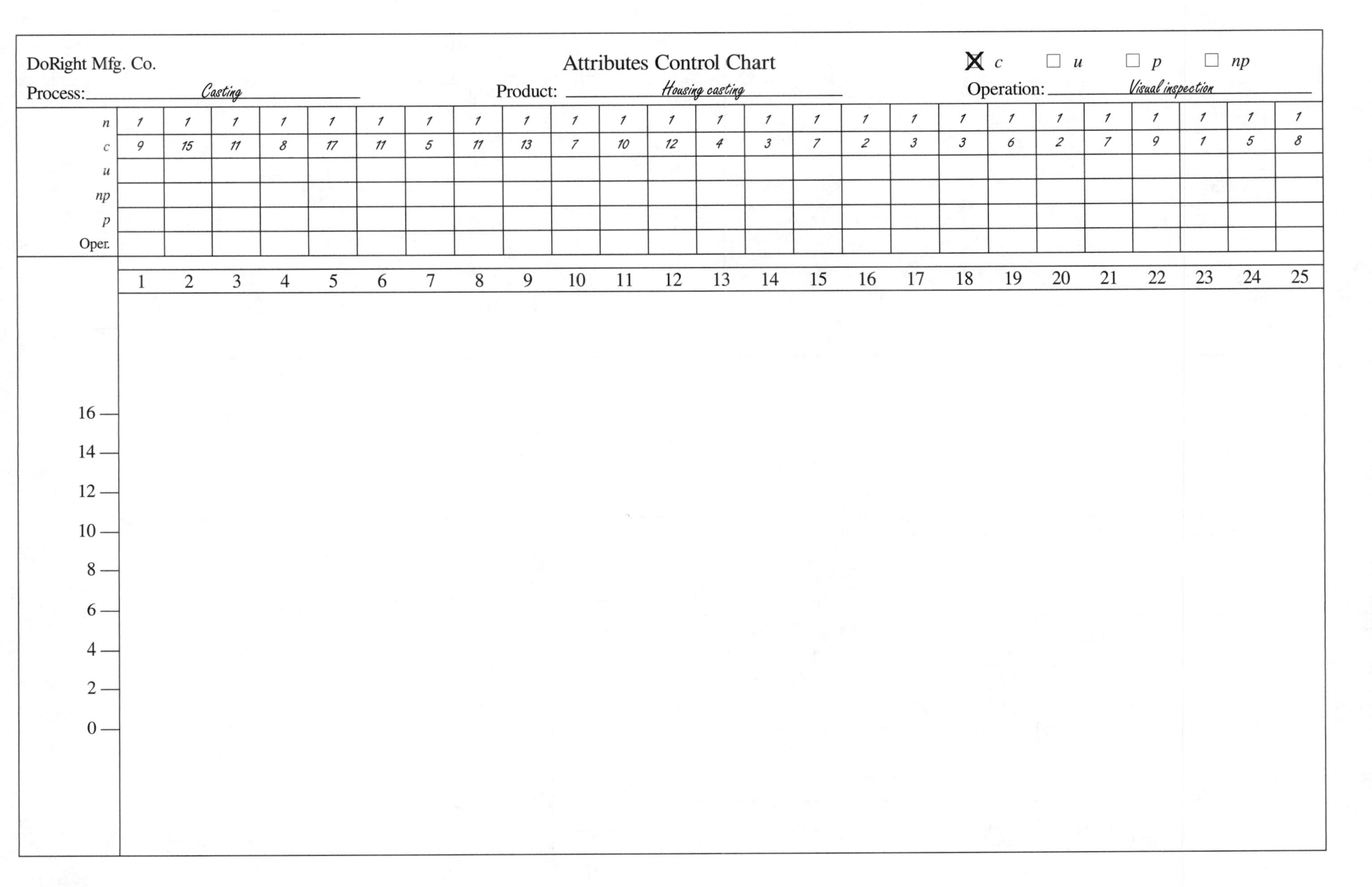

DoRight Mfg. Co.

Attributes Control Chart

☒ c ☐ u ☐ p ☐ np

Process: Casting   Product: Housing casting   Operation: Visual inspection

	1	2	3	4	5	6	7	8	9	10	11	12	13	14	15	16	17	18	19	20	21	22	23	24	25
n	1	1	1	1	1	1	1	1	1	1	1	1	1	1	1	1	1	1	1	1	1	1	1	1	1
c	9	15	11	8	17	11	5	11	13	7	10	12	4	3	7	2	3	3	6	2	7	9	1	5	8
u																									
np																									
p																									
Oper.																									

Figure 4.11. Practice c chart.

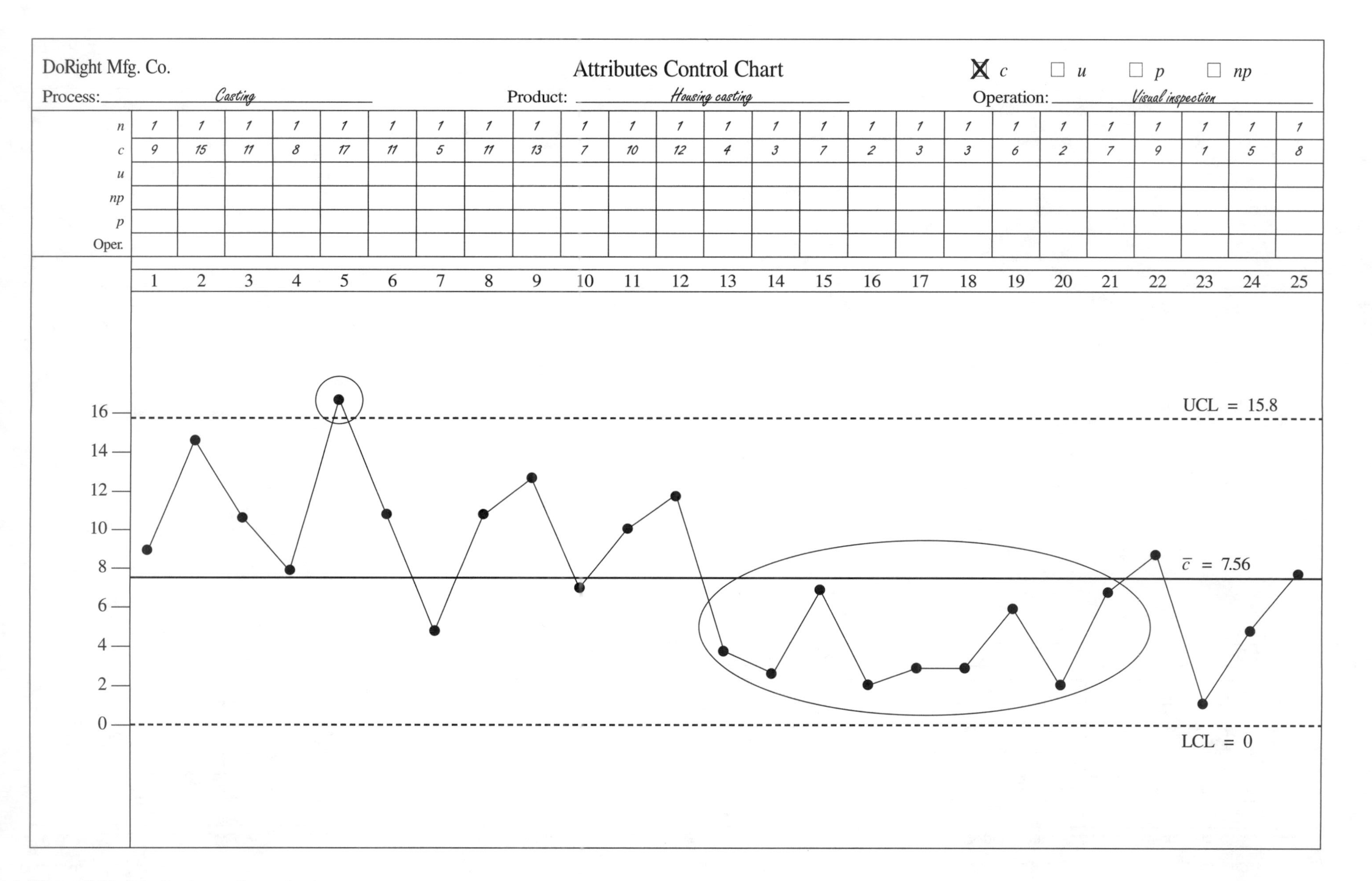

	1	2	3	4	5	6	7	8	9	10	11	12	13	14	15	16	17	18	19	20	21	22	23	24	25
n	1	1	1	1	1	1	1	1	1	1	1	1	1	1	1	1	1	1	1	1	1	1	1	1	1
c	9	15	11	8	17	11	5	11	13	7	10	12	4	3	7	2	3	3	6	2	7	9	1	5	8
u																									
np																									
p																									
Oper.																									

Figure 4.12. Solution to practice c chart.

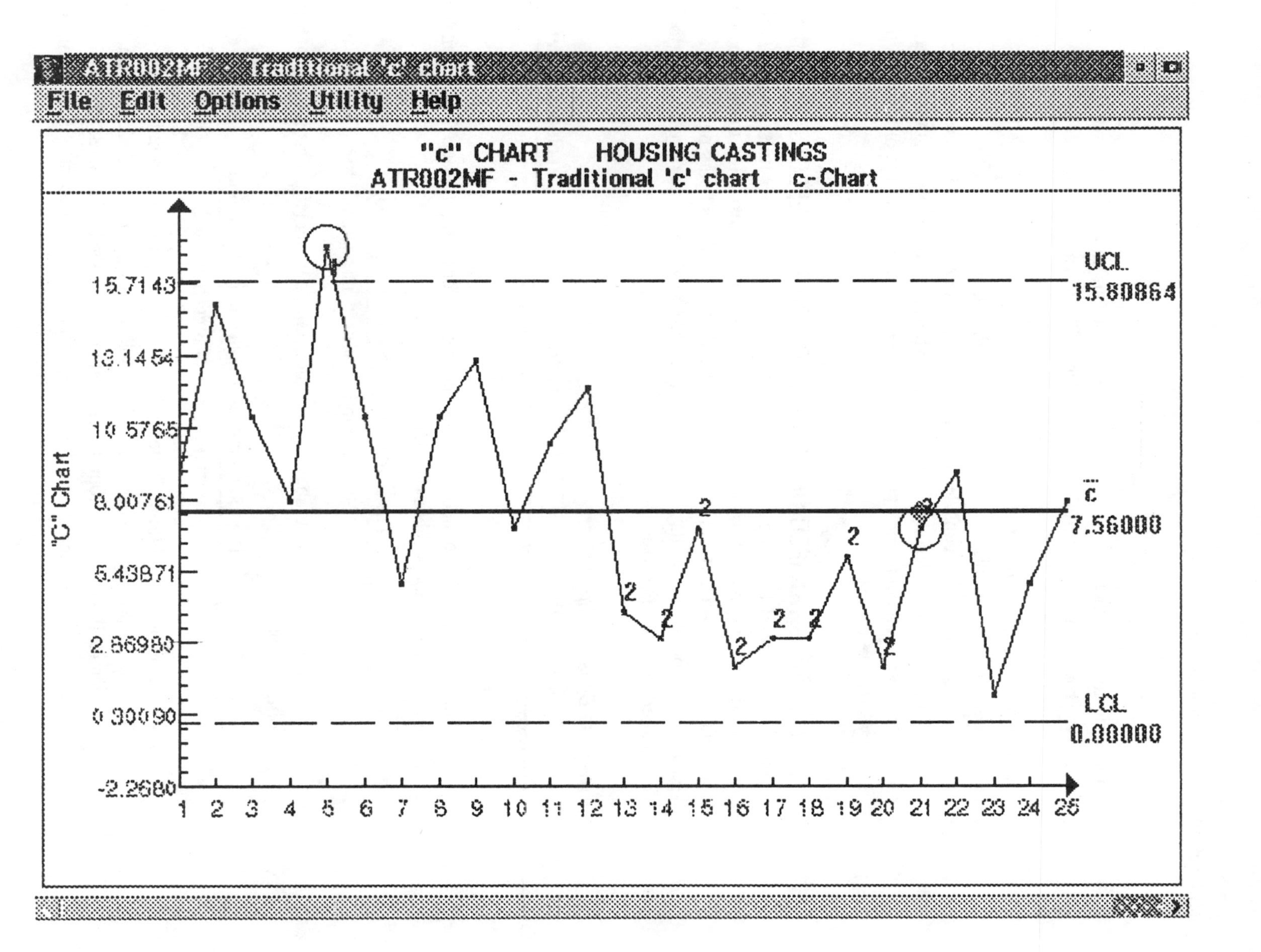

Courtesy of CIM Vision International

Figure 4.13. SQM software example of a *c* chart.

nonconformities, which are the largest contributors, can be corrected, which will cause major improvement in the overall process. Refer to Pareto analysis (chapter 10).

How to Interpret the c *Chart*

When determining control, the c chart is interpreted in the same way as the p chart, except nonconformities are being considered instead of nonconformances. The process capability is the average number of nonconformities ($\bar{c}$) when the process is in statistical control. Note: The terms *nonconformity* and *nonconformance* coincide with the older terms *defect* and *defective,* respectively.

THE *u* CHART—NUMBER OF NONCONFORMITIES (DEFECTS) PER UNIT

The u chart is used to control the number of nonconformities per unit. The u chart is concerned about the number of nonconformities that might be expected to be found on a per unit basis. The u chart can be applied to such diverse areas as hours per job, miles per gallon, jobs completed per day, and so on. It can be applied to virtually anything being considered on a per unit basis.

Steps to Construct the u *Chart*

Refer to Figures 4.14, 4.15, and 4.16 for practice with constant sample sizes. Refer to Figures 4.17, 4.18, and 4.19 for practice with varying sample sizes.

1. Collect the data.
2. Find and record the nonconformities per unit (u) for each subgroup. Note: Units can mean a part, subassembly, or assembly.

$$u = \frac{\text{The number of nonconformities found}}{\text{The total number of units inspected}}$$

3. Calculate the average number of nonconformities per unit ($\bar{u}$). This is the centerline of the u chart.

$$\bar{u} = \frac{\text{The total number of nonconformities found}}{\text{The total number of units inspected}}$$

4. Calculate the control limits. Note: If the LCL is a negative number, there is no LCL. The same rules apply to the u chart if sample sizes vary more than 25 percent; there are control limits plotted around each plotted point. (See Figure 4.18.) If sample sizes vary less than 25 percent, use the average sample size in the control limit formula.

$$\bar{u} \pm 3\left(\frac{\sqrt{\bar{u}}}{\sqrt{n}}\right)$$

5. Plot the centerline ($\bar{u}$) with a solid line and the control limits (UCL, LCL) with dashed lines after selecting the scale for the chart.
6. Plot the points (u) and connect them.
7. Interpret for control.

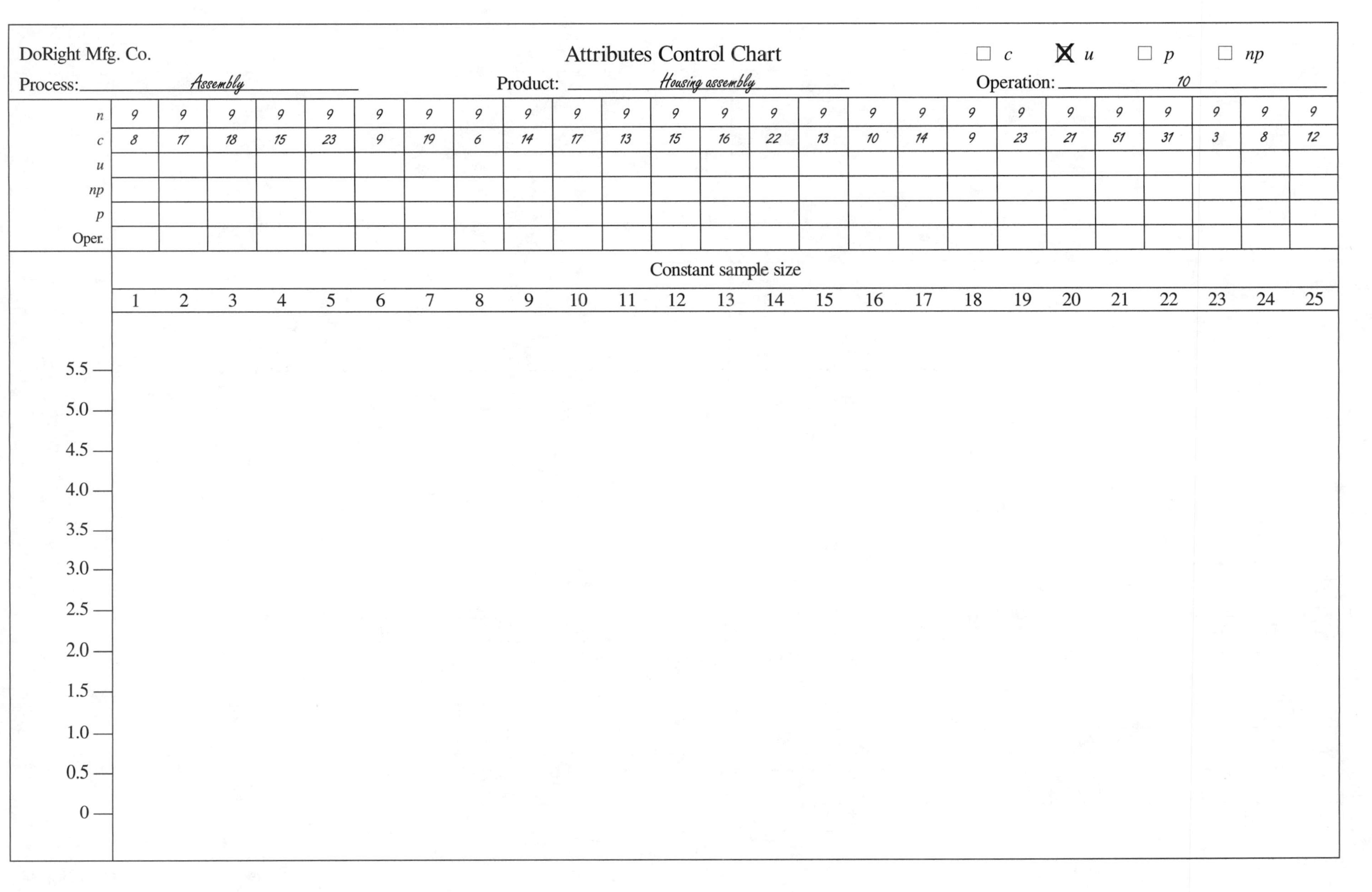

DoRight Mfg. Co.

Attributes Control Chart

☐ c ☒ u ☐ p ☐ np

Process: Assembly Product: Housing assembly Operation: 10

	1	2	3	4	5	6	7	8	9	10	11	12	13	14	15	16	17	18	19	20	21	22	23	24	25
n	9	9	9	9	9	9	9	9	9	9	9	9	9	9	9	9	9	9	9	9	9	9	9	9	9
c	8	17	18	15	23	9	19	6	14	17	13	15	16	22	13	10	14	9	23	21	51	31	3	8	12
u																									
np																									
p																									
Oper.																									

Figure 4.14. Practice u chart with constant sample sizes.

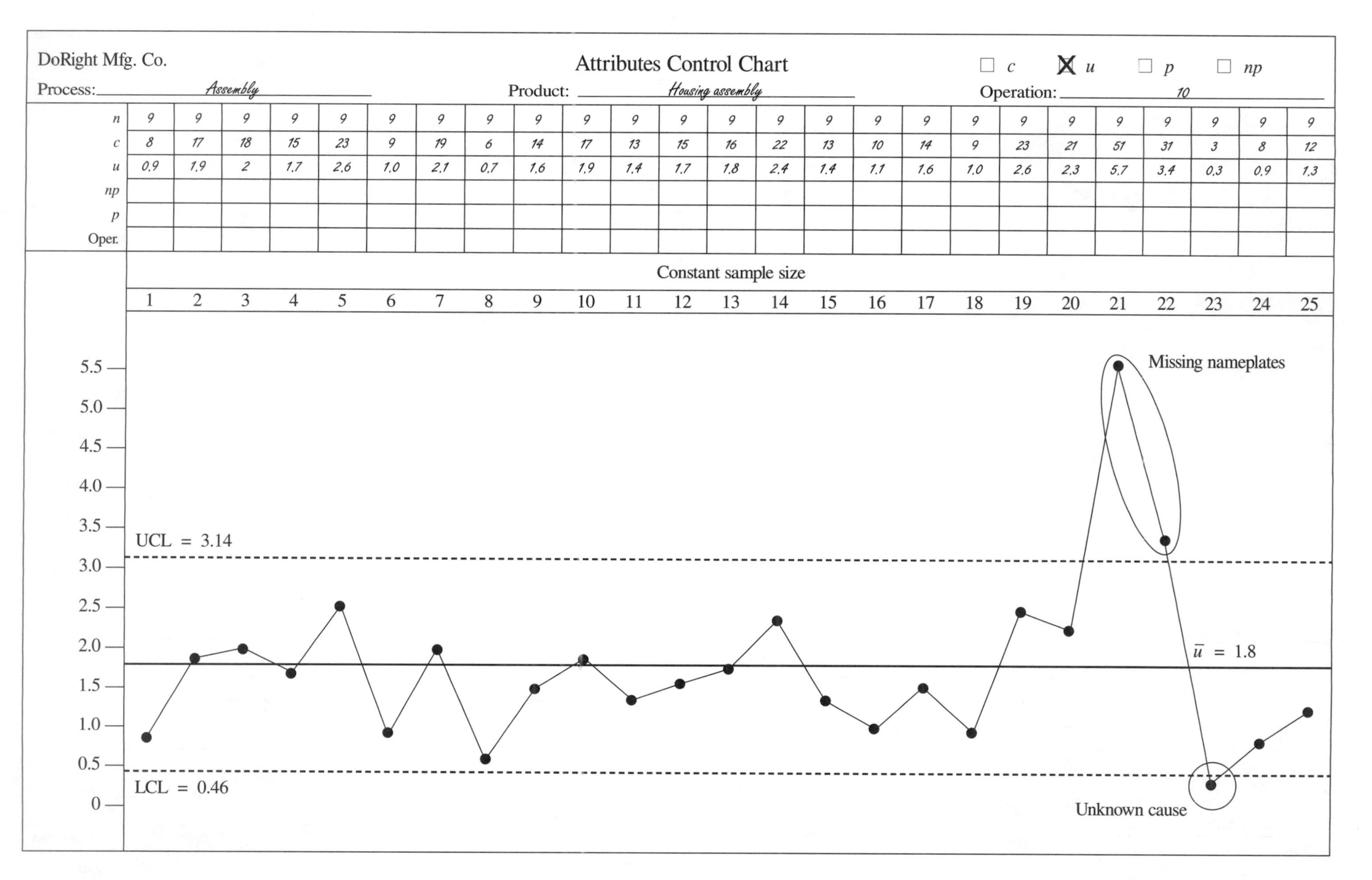

DoRight Mfg. Co.

Attributes Control Chart

☐ c ☒ u ☐ p ☐ np

Process: *Assembly* Product: *Housing assembly* Operation: *10*

	1	2	3	4	5	6	7	8	9	10	11	12	13	14	15	16	17	18	19	20	21	22	23	24	25
n	9	9	9	9	9	9	9	9	9	9	9	9	9	9	9	9	9	9	9	9	9	9	9	9	9
c	8	17	18	15	23	9	19	6	14	17	13	15	16	22	13	10	14	9	23	21	51	31	3	8	12
u	0.9	1.9	2	1.7	2.6	1.0	2.1	0.7	1.6	1.9	1.4	1.7	1.8	2.4	1.4	1.1	1.6	1.0	2.6	2.3	5.7	3.4	0.3	0.9	1.3
np																									
p																									
Oper.																									

Figure 4.15. Solution to practice *u* chart with constant sample sizes.

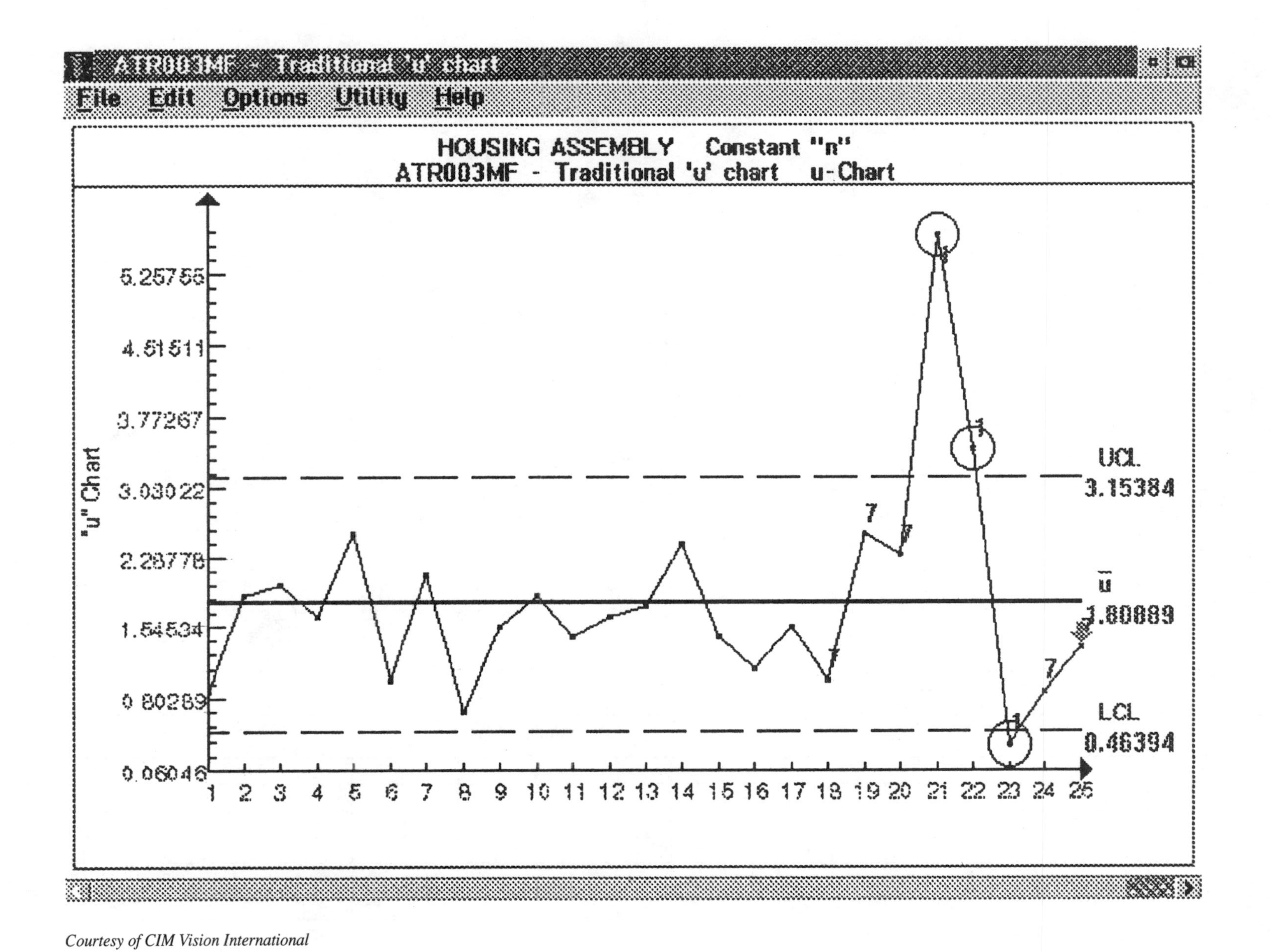

Courtesy of CIM Vision International

Figure 4.16. SQM software example of a *u* chart with constant sample sizes.

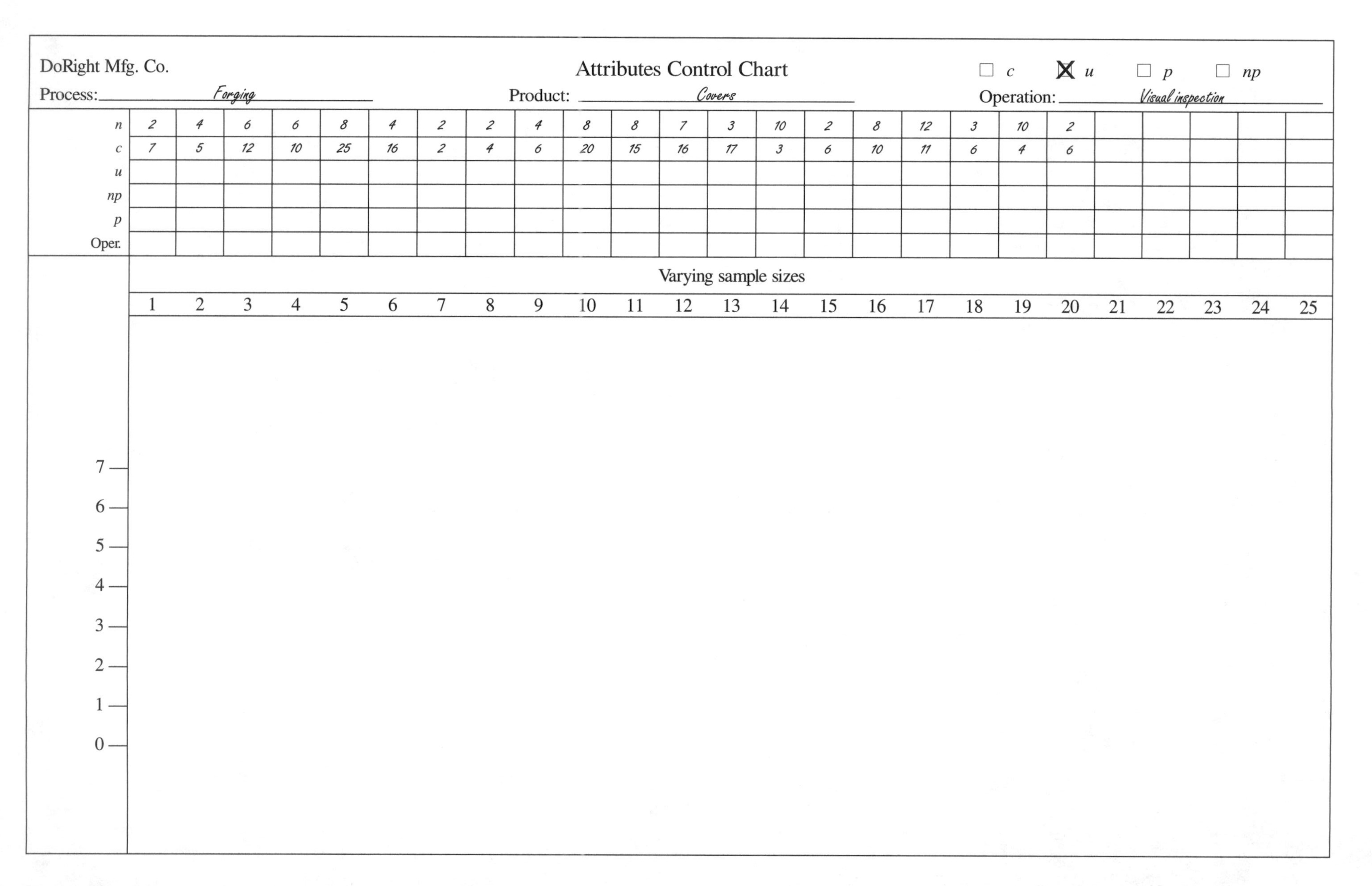

DoRight Mfg. Co.

Attributes Control Chart

☐ c ☒ u ☐ p ☐ np

Process: *Forging* Product: *Covers* Operation: *Visual inspection*

n	2	4	6	6	8	4	2	2	4	8	8	7	3	10	2	8	12	3	10	2					
c	7	5	12	10	25	16	2	4	6	20	15	16	17	3	6	10	11	6	4	6					
u																									
np																									
p																									
Oper.																									

Varying sample sizes

1	2	3	4	5	6	7	8	9	10	11	12	13	14	15	16	17	18	19	20	21	22	23	24	25

7
6
5
4
3
2
1
0

Figure 4.17. Practice u chart with varying sample sizes.

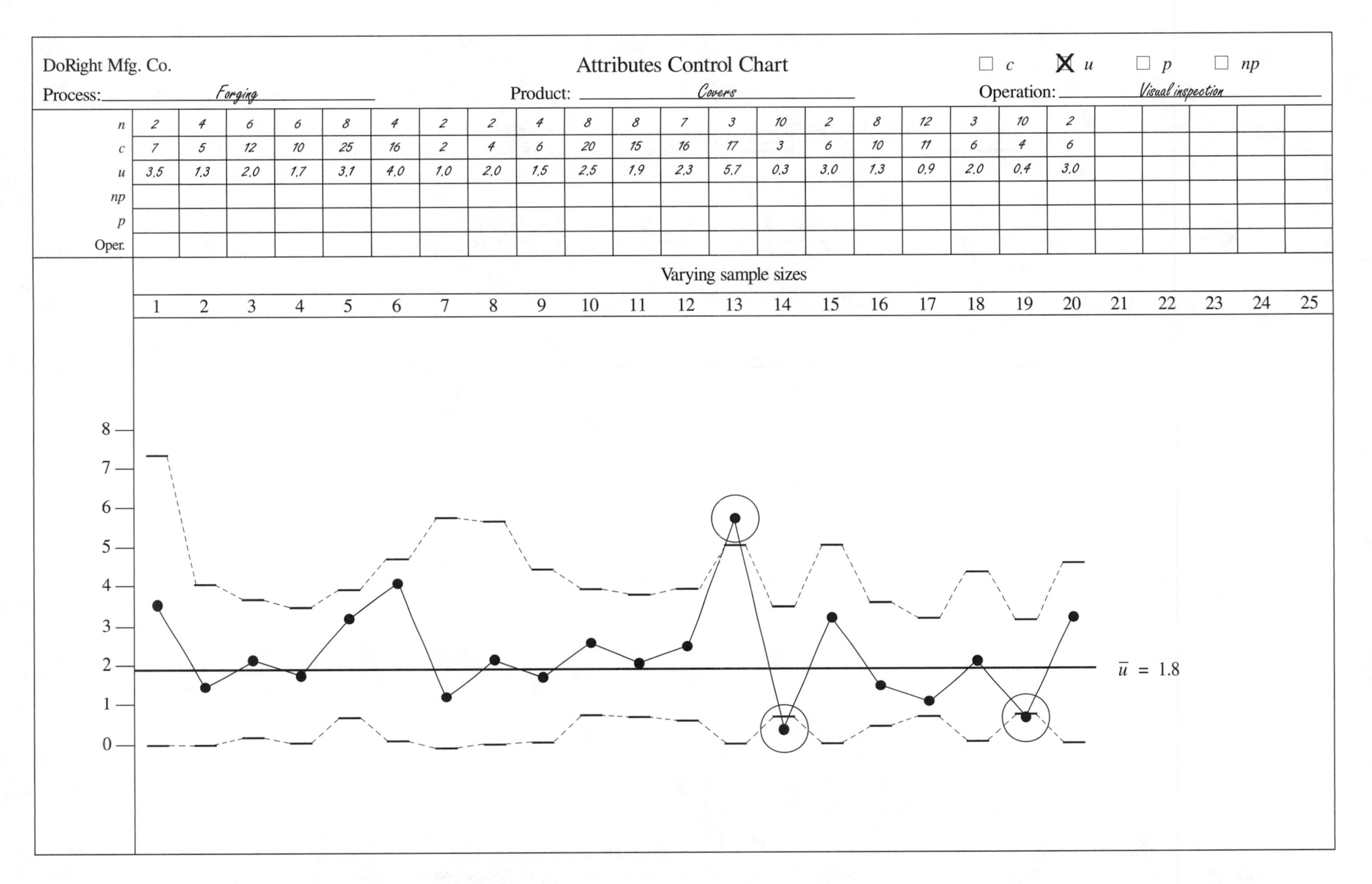

Figure 4.18. Solution to practice u chart with varying sample sizes.

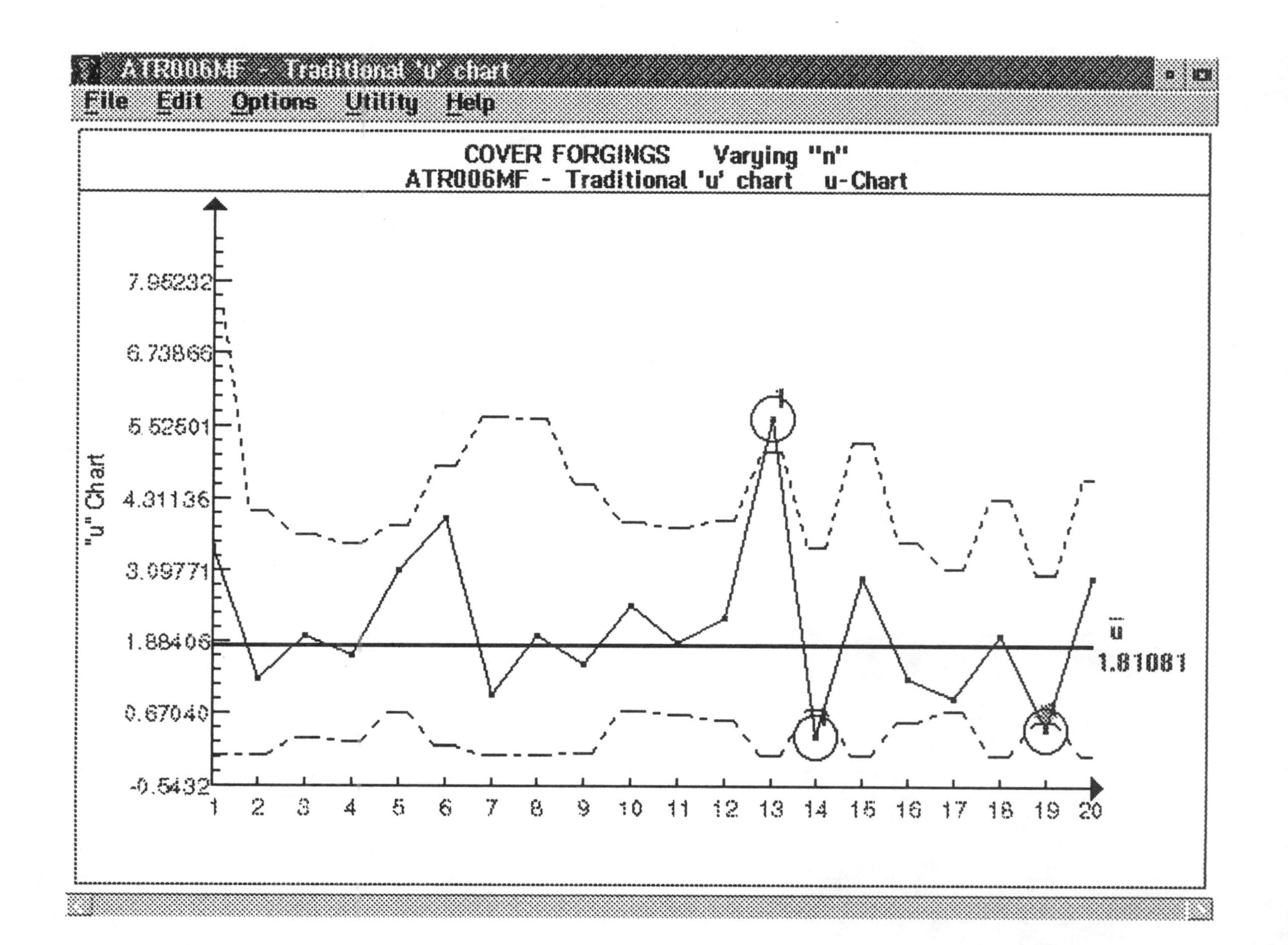

Courtesy of CIM Vision International

Figure 4.19. SQM software example of a *u* chart with varying sample sizes.

The u chart is similar to interpreting the c chart except that nonconformities per unit are being considered. The capability of the process is the average ($\bar{u}$) if the process is in statistical control.

PROCESS CAPABILITY—ATTRIBUTE CHARTS

The primary indicators of process capability with attributes are

Percent yield (percent yield)

$$\text{Percent}_{\text{yield}} = 100 \text{ percent} - \text{Percent}_{\text{defective}}$$

Parts per million defective (PPM)

$$\text{PPM} = \text{Percent}_{\text{defective}} \times 10{,}000$$

The *percent yield of a process* is defined as the percentage of good parts that are to be expected from the process the first time through. When a p chart is used and the process is in statistical control, the average percent defective represents the ongoing level of poor performance for the process. Therefore, the percent yield of the process is 100 percent minus the percent defective. If a process has an average fraction defective of .05, the percent defective equals 5 percent, and the percent yield equals 95 percent.

Where np charts (for number of defectives) are used, one must first divide the average number of defectives by the production quantity times 100 to obtain the percent defective. Then subtract the percent defective from 100 percent to obtain the percent yield. Where c and u charts are concerned, translate the defects data into percent defective, then compute the percent yield.

Another popular method for monitoring the output of a process is parts per million (PPM). PPM computation is focused on the defective output of the process.

For example, say a process is being monitored using a p chart, and the process is in statistical control. The process average (from the p chart) is .02 (or 2 percent defective). Therefore, the percent yield is equal to 100 percent minus 2 percent or 98 percent good product. The PPM is equal to 2 percent times 10,000 or 20,000 parts per million defective. The intention of PPM, generally, is to explode the value of percent defective so that it does not appear to be acceptable. For example, 0.05 percent defective looks small and gives one the impression of being acceptable. Today, the old idea of acceptable defectives is going out of style and is being replaced by goals and methods for defect/defective prevention. The PPM associated with 0.05 percent is 500 parts per million.

REVIEW PROBLEMS

Refer to Appendix H for answers.

1. Find the fraction nonconforming (p) for the following data.

   Sample (n) = 80 Defectives (np) = 12

2. Find the average fraction nonconforming ($\bar{p}$) using the following data.

Samples (n)	75	82	79	71	81
Nonconformances (np)	3	7	12	3	13

3. Find the UCL for a p chart using the following information.

   $\bar{p} = 0.10 \qquad n = 30$

4. The following number of nonconformities were found. Find the average ($\bar{c}$).

   25 12 17 3 19

5. Find the UCL for a c chart if the average ($\bar{c}$) is 9.
6. Find the LCL for a p chart if $n = 50$ and $\bar{p} = .05$.
7. Find the LCL for a c chart if $\bar{c} = 13$.
8. If you inspect 50 parts and find 12 nonconformities in all, what is the number of nonconformities per unit?
9. Find the average number of nonconformities per unit ($\bar{u}$) using the following data.

Nonconformities found	13	19	5	17	2
Units inspected	5	5	5	5	5

10. Find the UCL for a u chart if $\bar{u} = 12$ and $n = 5$.
11. If a p chart shows a downward trend toward zero, you should
    a. Do nothing, it's good news
    b. Circle the trend
    c. Take action to find the cause
    d. None of the above
12. Which of the following control charts could be used to control the number of typing errors per memo?
    a. c chart
    b. p chart
    c. u chart
    d. R chart
13. Which of the following techniques is used to identify the most significant problem in an area?
    a. Normal distribution
    b. Control chart
    c. Pareto analysis
    d. None of the above
14. When using p and u charts, if the sample sizes are not constant, but close, you should
    a. Not use the chart
    b. Use the average sample size in the formula
    c. Scatter the points
    d. Use a different chart scale

15. Which of the following charts is used to control the quantity nonconforming?

    a. *u* chart

    b. *c* chart

    c. *p* chart

    d. *np* chart

16. Causes of problems that tend to occur only once (isolated cases) are called ________________ problems.

17. Improvement, using attribute charts, is more effective if ________________ analysis is used to isolate the sources of problems that support the chronic process average.

18. A freak (one plotted point beyond a control limit) on an attribute chart indicates that the cause is (sporadic or chronic).

19. More than one Pareto analysis is often necessary on a process so that the sources of a problem are more clearly understood. (True or false)

20. Which of the following cases causes control limits to vary around each plotted point?

    a. Defects vary significantly

    b. Sample sizes vary significantly

    c. The process varies erratically

    d. Many adjustments are made on the process

21. A process is in control using a *p* chart for proportion defective. The process average is .025.

    a. What is the percent yield of the process?

    b. What is the number of defectives stated in terms of PPM?

CHAPTER 5

Short-Run Attribute Control Charts

INTRODUCTION

All of the traditional attribute control charts become short-run charts by using the standardized approach. The application for short-run attribute charts is, specifically, when one chooses to chart different processes (with different process averages) on the same chart. For example, it has been decided that four processes will be charted using a p chart for proportion defective. Table 5.1 shows the four different processes and their process averages ($\bar{p}$).

If one were to plot these four different processes on a traditional p chart, there would be interpretation problems because each time that data were taken from a different process, the chart would appear to go out of control (or in control) when it is not. The only way to avoid false signals of process control is to plot the four different processes on a short-run standardized p chart (or choose to plot an attribute chart for each of the four processes).

STANDARDIZED CHARTS

Standardized charts, in effect, allow the combination of different process averages onto one chart because each plot point (from each different process) is plotted by calculating the difference between the latest plot point value from process A and the historical process average of process A divided by the historical process A standard deviation. In other words, the plot point values always reflect the proportion and direction of historical control limits consumed by the most recent data. Standardized charts offer a statistically sound method of mixing different processes (with different process averages) onto the same chart, but they

Table 5.1. Process averages.

Process	Process average
A	$\bar{p}$ = .05
B	$\bar{p}$ = .01
C	$\bar{p}$ = .12
D	$\bar{p}$ = .20

are more cumbersome to use. More likely then not, users tend to use the traditional attribute chart on each process instead of a standardized chart.

SHORT-RUN STANDARDIZED *p* CHART

The applications for the short-run standardized p chart are the same as traditional p charts, except it is assumed that the user wants to plot the proportion defective from different processes with different process averages on the same chart. Once again, as with all standardized charting methods, the centerline and control limits are fixed. The centerline is always zero, and the upper and lower control limits are fixed at plus and minus three, respectively. These limits never need to be recalculated (as do traditional p charts).

The plot points must be standardized with respect to their historical values before they are plotted.

Using the standardized p chart, the proportion defective (p) must be computed for the recent subgroup from the process, then that p value must be standardized prior to plotting. The p value for the most recent subgroup, as covered in traditional p charts is

$$p = \frac{\Sigma \text{ of defectives}}{\Sigma \text{ of samples}}$$

Next, the p value must be standardized with respect to the historical $\bar{p}$ value from that process before it can be plotted.

$$p_{plot} = \frac{P_A - \bar{P}_{HA}}{\sigma_{P_{HA}}}$$

where

p_A = the computed p value for the recent subgroup from process A
$\bar{p}_{HA}$ = the historical process average from process A
$\sigma_{P_{HA}}$ = the historical standard deviation of p from process A

$$\sigma_{P_{HA}} = \sqrt{\frac{\bar{p}\,(1 - \bar{p})}{n}}$$

where

$\bar{p}$ = the updated historical process average from process A
n = the constant sample sizes taken from process A

Also, historical values for each process must be updated when the process is out of control and assignable causes have not been corrected. If the process remains in statistical control or the assignable cause for out-of-control points has been isolated and corrected, the historical values need not be updated. This general rule applies to all standardized charts. For practice in plotting the standardized p chart, refer to Figures 5.1 and 5.2.

SHORT-RUN STANDARDIZED *c* CHARTS

The applications for the short-run standardized c chart are same as with traditional c charts, except it is assumed that the user wants to plot the number of defects from different processes on the same chart. Once again, as with all standardized charting methods, the centerline and control limits are fixed. The centerline is always zero, and the upper and

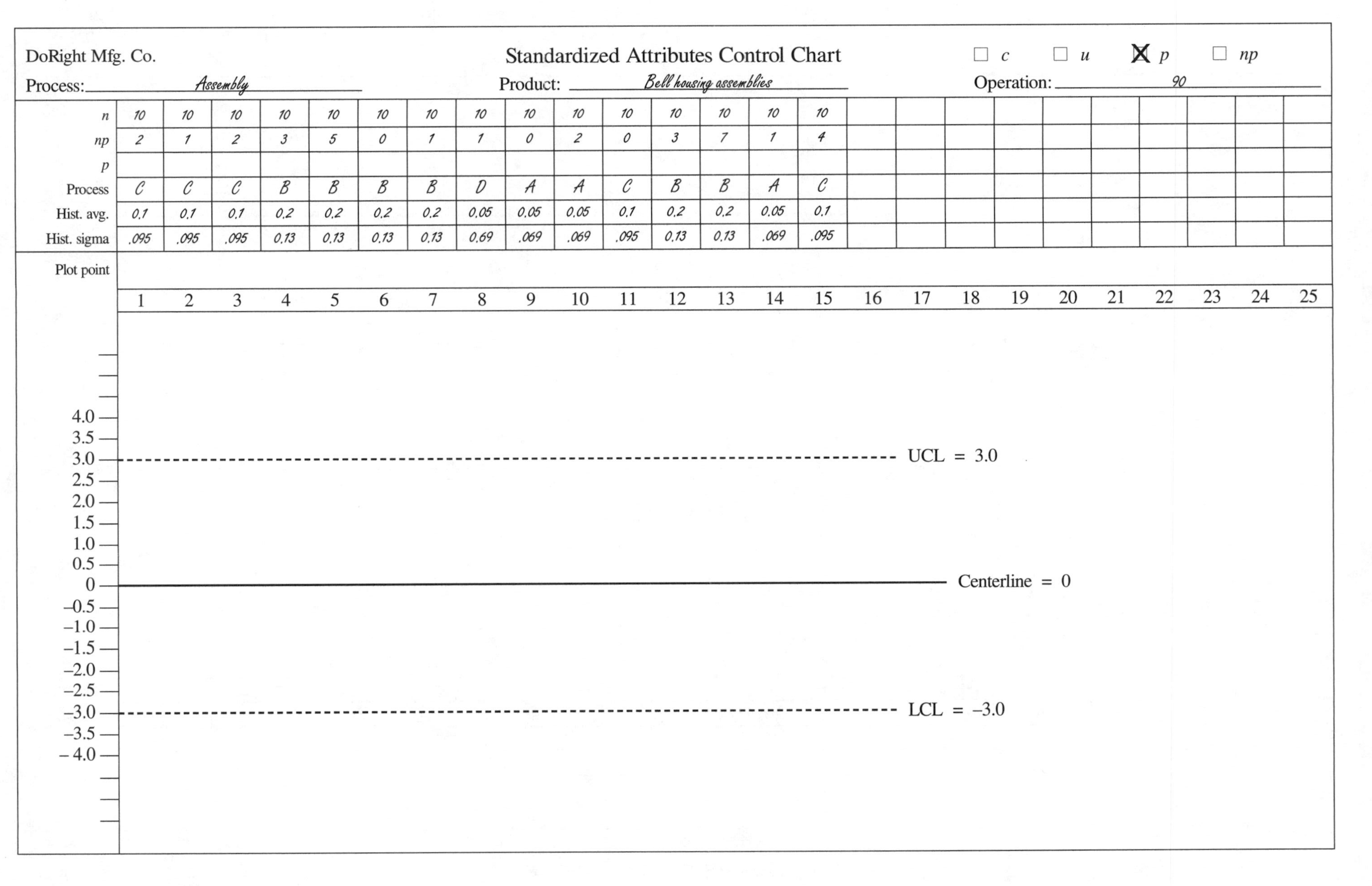

DoRight Mfg. Co.

Standardized Attributes Control Chart

☐ c ☐ u ☒ p ☐ np

Process: Assembly

Product: Bell housing assemblies

Operation: 90

n	10	10	10	10	10	10	10	10	10	10	10	10	10	10	10										
np	2	1	2	3	5	0	1	1	0	2	0	3	7	1	4										
p																									
Process	C	C	C	B	B	B	B	D	A	A	C	B	B	A	C										
Hist. avg.	0.1	0.1	0.1	0.2	0.2	0.2	0.2	0.05	0.05	0.05	0.1	0.2	0.2	0.05	0.1										
Hist. sigma	.095	.095	.095	0.13	0.13	0.13	0.13	0.69	.069	.069	.095	0.13	0.13	.069	.095										
Plot point																									
	1	2	3	4	5	6	7	8	9	10	11	12	13	14	15	16	17	18	19	20	21	22	23	24	25

Figure 5.1. Practice standardized *p* chart.

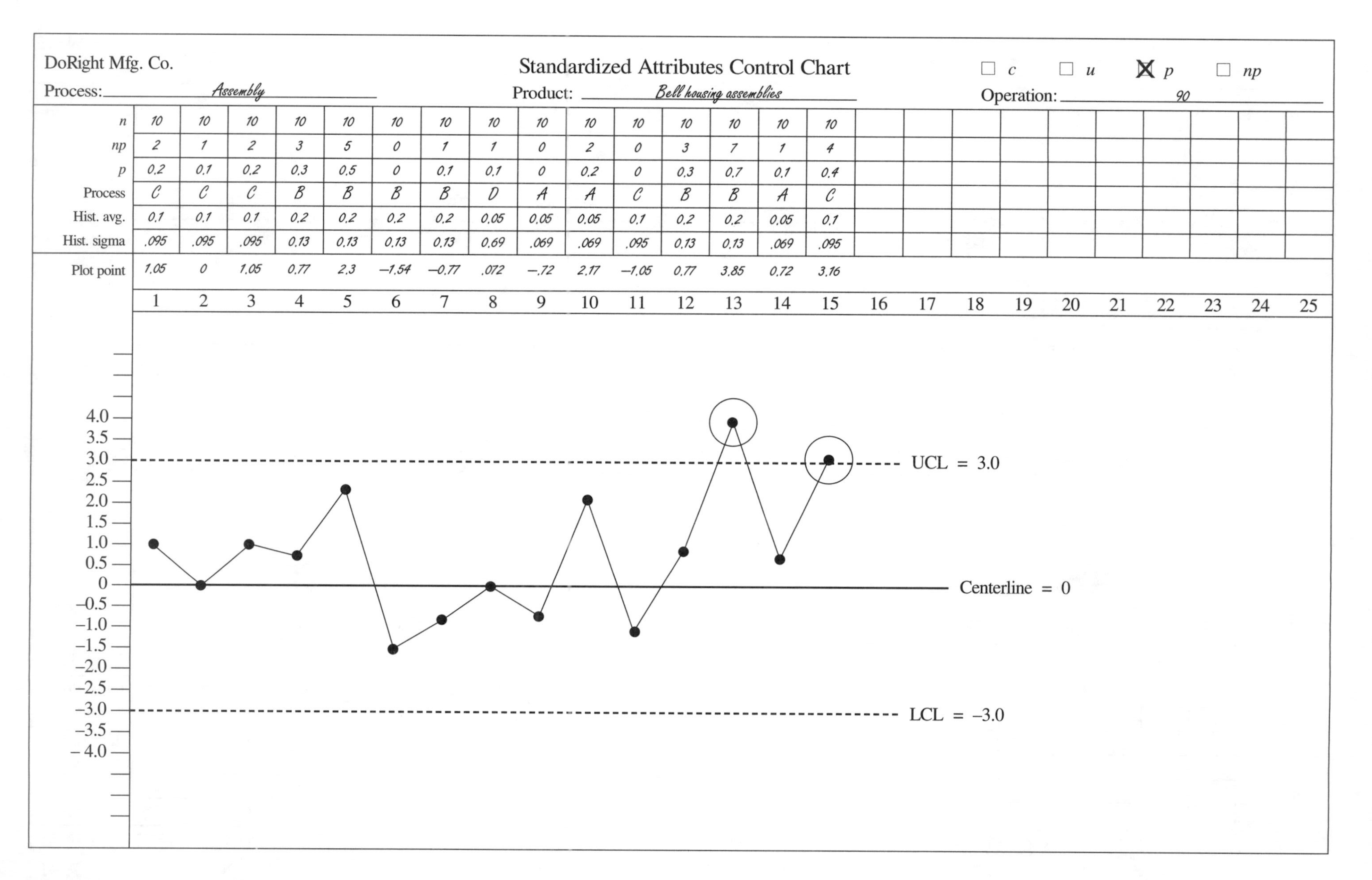

DoRight Mfg. Co.

Standardized Attributes Control Chart

☐ c ☐ u ☒ p ☐ np

Process: Assembly

Product: Bell housing assemblies

Operation: 90

	1	2	3	4	5	6	7	8	9	10	11	12	13	14	15	16	17	18	19	20	21	22	23	24	25
n	10	10	10	10	10	10	10	10	10	10	10	10	10	10	10										
np	2	1	2	3	5	0	1	1	0	2	0	3	7	1	4										
p	0.2	0.1	0.2	0.3	0.5	0	0.1	0.1	0	0.2	0	0.3	0.7	0.1	0.4										
Process	C	C	C	B	B	B	B	D	A	A	C	B	B	A	C										
Hist. avg.	0.1	0.1	0.1	0.2	0.2	0.2	0.2	0.05	0.05	0.05	0.1	0.2	0.2	0.05	0.1										
Hist. sigma	.095	.095	.095	0.13	0.13	0.13	0.13	0.69	.069	.069	.095	0.13	0.13	.069	.095										
Plot point	1.05	0	1.05	0.77	2.3	–1.54	–0.77	.072	–.72	2.17	–1.05	0.77	3.85	0.72	3.16										

Figure 5.2. Solution to practice standardized *p* chart.

lower control limits are fixed at plus and minus 3, respectively. These limits never need to be recalculated (as do traditional c charts).

The plot points must be standardized with respect to their historical values before they are plotted.

Using the standardized c chart, the number of defects (c) must be computed for the recent subgroup from the process, then that c value must be standardized prior to plotting. The c value for the most recent subgroup, as covered in traditional c charts is the number of defects found in that subgroup. Next, the c value must be standardized with respect to the historical $\bar{c}$ value from that process before it can be plotted.

$$C_{plot} = \frac{C_A - \bar{C}_{HA}}{\sigma_{C_{HA}}}$$

where

c_A = the number of defects c for the recent subgroup from process A
$\bar{C}_{HA}$ = the historical process average of defects from process A
$\sigma_{C_{HA}}$ = the historical standard deviation of c from process A

$$\sigma_{C_{HA}} = \sqrt{\bar{C}_{HA}}$$

where

$\bar{C}_{HA}$ = the updated historical process average from process A

Also, remember that the historical values must be updated properly for this chart to be valid. For practice in plotting the standardized c chart, refer to Figures 5.3 and 5.4.

SHORT-RUN STANDARDIZED *u* CHARTS

The applications for the short-run standardized u chart are same as with traditional u charts, except it is assumed that the user wants to plot the number of defects per unit from different processes on the same chart. Once again, as with all standardized charting methods, the centerline and control limits are fixed. The centerline is always zero, and the upper and lower control limits are fixed at plus and minus three respectively. These limits never need to be recalculated (as do traditional u charts).

The plot points must be standardized with respect to their historical values before they are plotted.

Using the standardized u chart, the number of defects per unit (u) must be computed for the recent subgroup from the process, then that u value must be standardized prior to plotting. The u value for the most recent subgroup, as covered in traditional u charts, is the number of defects found in that subgroup.

Next, the u value must be standardized with respect to the historical $\bar{u}$ value from that process before it can be plotted.

$$u_{plot} = \frac{u_A - \bar{u}_{HA}}{\sigma_{U_{HA}}}$$

where

u_A = the number of defects per unit from process (or unit) A
$\bar{u}_{HA}$ = the historical process average of u from process (or unit) A
$\sigma_{u_{HA}}$ = the historical standard deviation of u from process (or unit) A

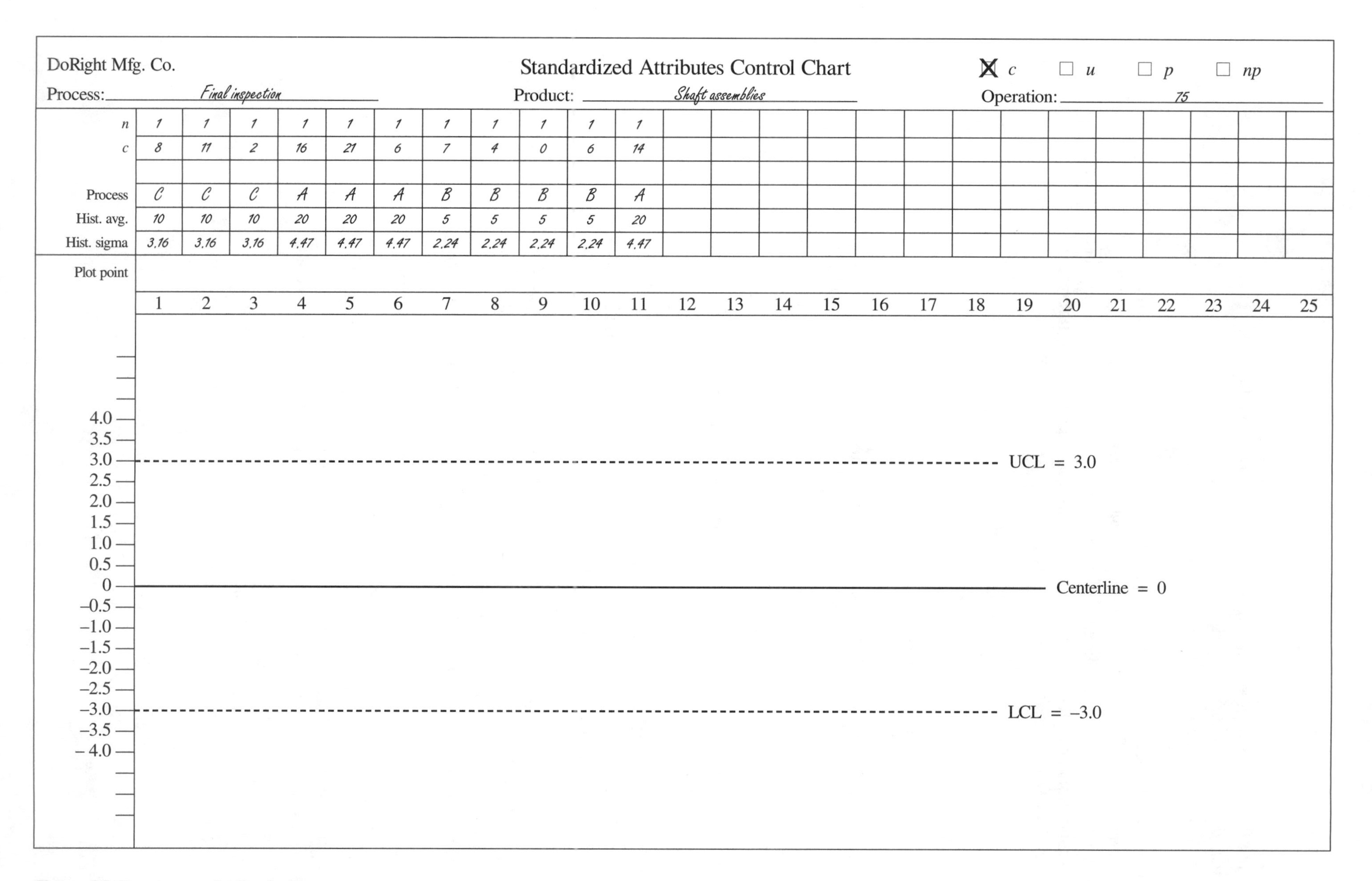

DoRight Mfg. Co.

Standardized Attributes Control Chart

☒ c ☐ u ☐ p ☐ np

Process: Final inspection

Product: Shaft assemblies

Operation: 75

	1	2	3	4	5	6	7	8	9	10	11	12	13	14	15	16	17	18	19	20	21	22	23	24	25
n	1	1	1	1	1	1	1	1	1	1	1														
c	8	11	2	16	21	6	7	4	0	6	14														
Process	C	C	C	A	A	A	B	B	B	B	A														
Hist. avg.	10	10	10	20	20	20	5	5	5	5	20														
Hist. sigma	3.16	3.16	3.16	4.47	4.47	4.47	2.24	2.24	2.24	2.24	4.47														
Plot point																									

Figure 5.3. Practice standardized c chart.

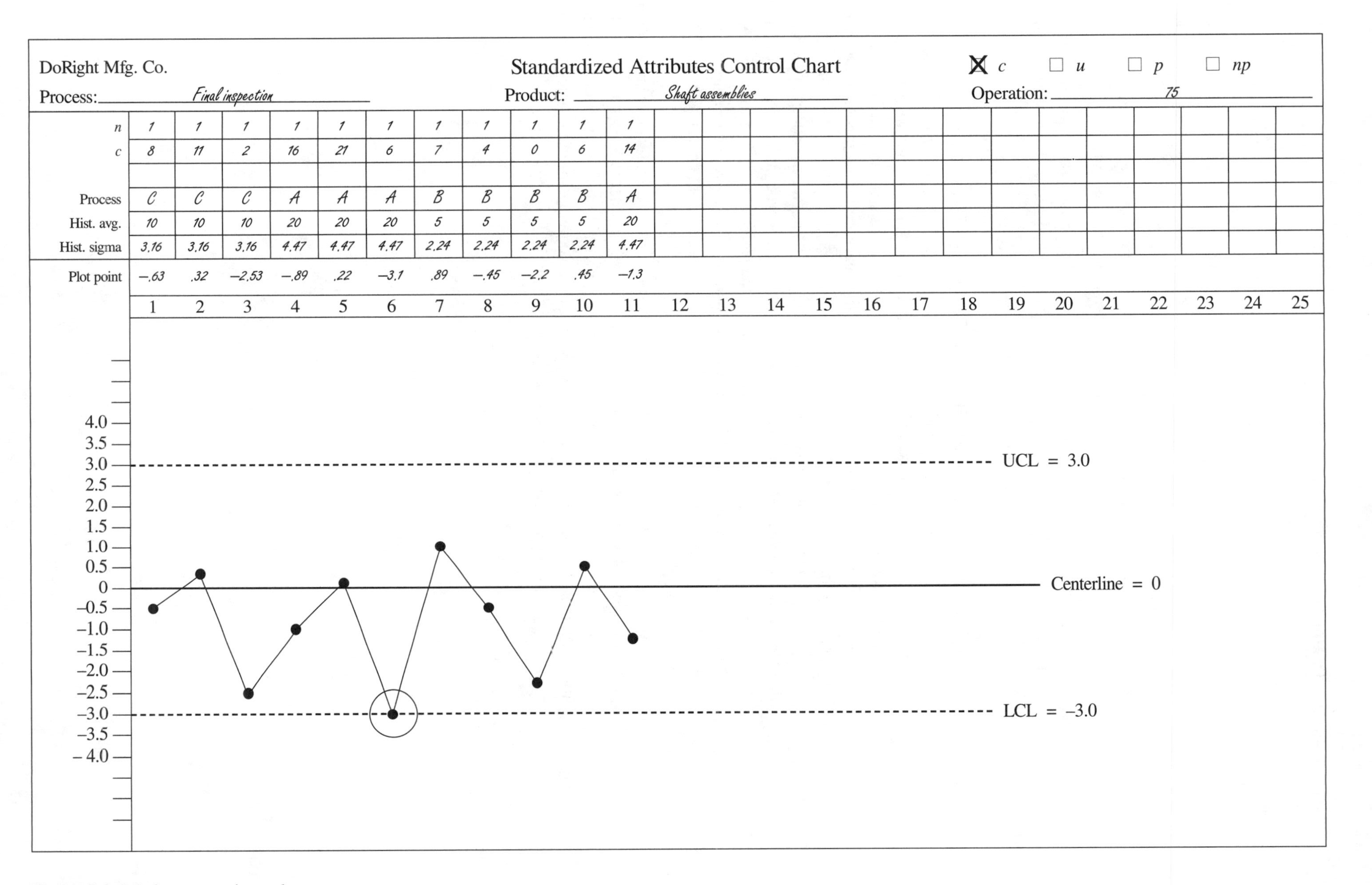

	1	2	3	4	5	6	7	8	9	10	11	12	13	14	15	16	17	18	19	20	21	22	23	24	25
n	1	1	1	1	1	1	1	1	1	1	1														
c	8	11	2	16	21	6	7	4	0	6	14														
Process	C	C	C	A	A	A	B	B	B	B	A														
Hist. avg.	10	10	10	20	20	20	5	5	5	5	20														
Hist. sigma	3.16	3.16	3.16	4.47	4.47	4.47	2.24	2.24	2.24	2.24	4.47														
Plot point	–.63	.32	–2.53	–.89	.22	–3.1	.89	–.45	–2.2	.45	–1.3														

Figure 5.4. Solution to practice *c* chart.

$$\sigma_{u_{HA}} = \sqrt{\frac{\bar{u}_{HA}}{n}}$$

where

$\bar{u}_{HA}$ = the updated historical process average from process (or unit) A

n = the constant sample size taken from process (or unit) A

Also, remember that the historical values must be properly updated for the standardized u chart to be valid. For practice in plotting the standardized u chart, refer to Figures 5.5 and 5.6.

INTERPRETING STANDARDIZED ATTRIBUTE CHARTS

Interpretation of control using standardized attribute charts is cumbersome at best. Just because the different plot points are between control limits, this does not mean that each of the different processes are in control. For example, review the standardized attribute chart in Figure 5.7. Notice that all of the plot points on the chart are between control limits and they appear to have no trends or runs.

Looking closer, however, you will find that there is a 12-point run pattern in process B that has been overlooked. A better idea, using standardized charts, may be to plot different processes with different types of symbols to help see the patterns of each individual process over time (such as the Xs used to show the run pattern of process B in Figure 5.8). Of course, if each of these processes were plotted on its own attribute chart, this would not have been a problem.

As you can see, the standardized attribute charts are very complex in terms of computing the plot point values, and updating historical values in order to keep the chart valid. I prefer to use traditional attribute charts on one process instead of the standardized approach in order to keep it simple on the processor. The coverage of these types of charts is only intended to give readers the option of using them, if they so desire.

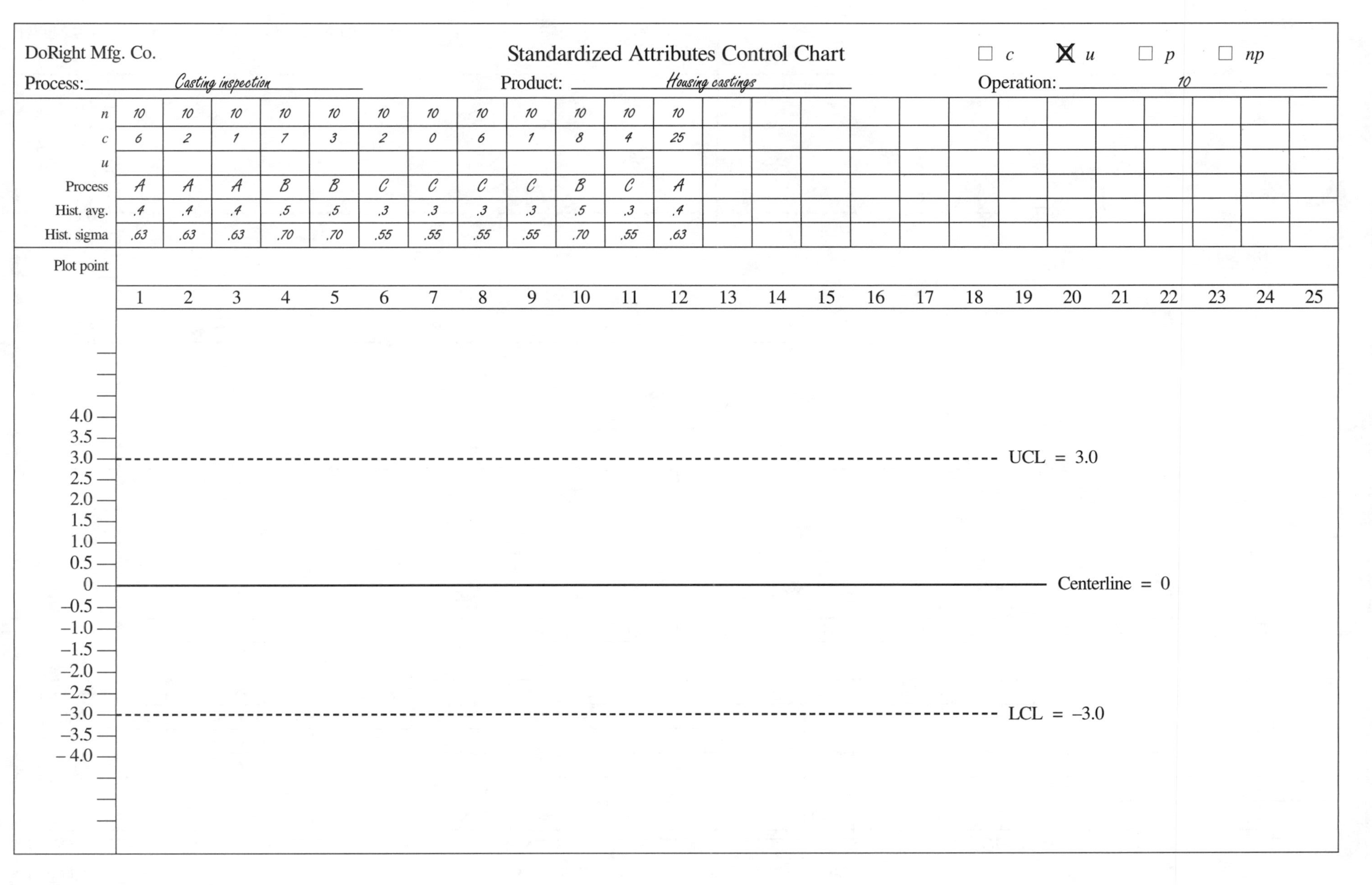
DoRight Mfg. Co.

Standardized Attributes Control Chart

☐ c ☒ u ☐ p ☐ np

Process: Casting inspection

Product: Housing castings

Operation: 10

	1	2	3	4	5	6	7	8	9	10	11	12	13	14	15	16	17	18	19	20	21	22	23	24	25
n	10	10	10	10	10	10	10	10	10	10	10	10													
c	6	2	1	7	3	2	0	6	1	8	4	25													
u																									
Process	A	A	A	B	B	C	C	C	C	B	C	A													
Hist. avg.	.4	.4	.4	.5	.5	.3	.3	.3	.3	.5	.3	.4													
Hist. sigma	.63	.63	.63	.70	.70	.55	.55	.55	.55	.70	.55	.63													
Plot point																									

Figure 5.5. Practice standardized *u* chart.

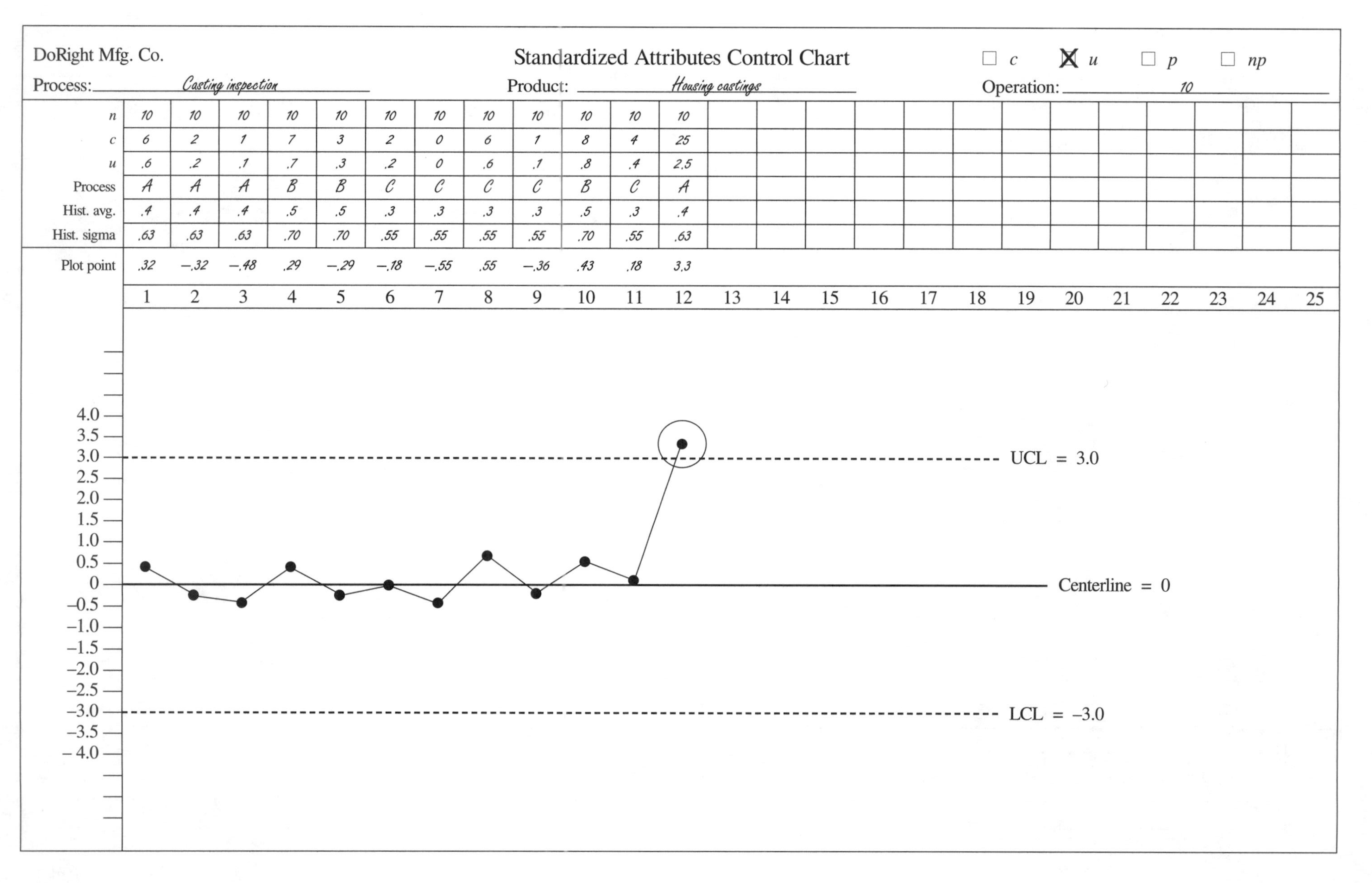
DoRight Mfg. Co.

Standardized Attributes Control Chart

☐ c ☒ u ☐ p ☐ np

Process: Casting inspection

Product: Housing castings

Operation: 10

	1	2	3	4	5	6	7	8	9	10	11	12	13	14	15	16	17	18	19	20	21	22	23	24	25
n	10	10	10	10	10	10	10	10	10	10	10	10													
c	6	2	1	7	3	2	0	6	1	8	4	25													
u	.6	.2	.1	.7	.3	.2	0	.6	.1	.8	.4	2.5													
Process	A	A	A	B	B	C	C	C	C	B	C	A													
Hist. avg.	.4	.4	.4	.5	.5	.3	.3	.3	.3	.5	.3	.4													
Hist. sigma	.63	.63	.63	.70	.70	.55	.55	.55	.55	.70	.55	.63													
Plot point	.32	–.32	–.48	.29	–.29	–.18	–.55	.55	–.36	.43	.18	3.3													

Figure 5.6. Solution to practice standardized *u* chart.

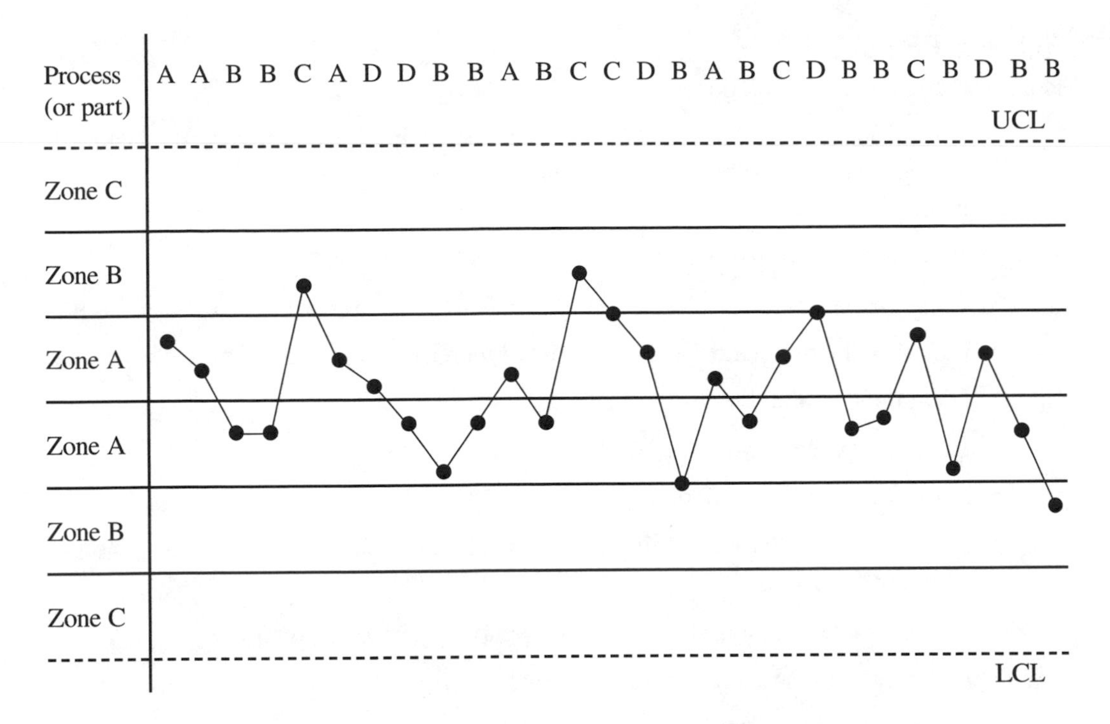

Figure 5.7. A standardized attribute chart, which appears to be in control.

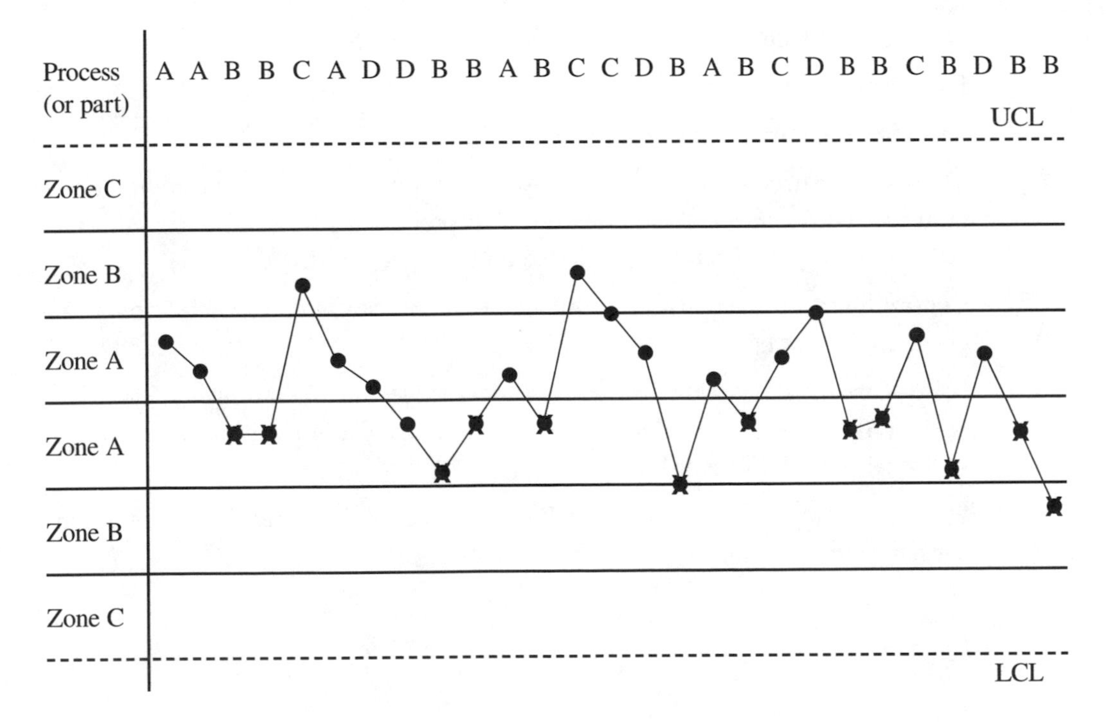

Figure 5.8. A standardized attribute chart, in which process B is out of control.

REVIEW PROBLEMS

Refer to Appendix H for solutions.

1. A method of coding data that factors each plot point for a part using the historical average and standard deviation for that part is called the ________________ method.
2. What is the centerline for all standardized attribute charts?
3. What are the upper and lower control limits for all standardized attribute charts?
4. Calculate the plot point for a standardized p chart using the following information.

   Computed p value for part A = .05

   Historical process average (part A) = .03

   Historical standard deviation (part A) = .022
5. A standardized c chart is being used. Calculate the plot point value using the following information.

   Number of defects found in recent subgroup for part B: 12

   Historical process average for part B = 15

   Historical standard deviation for part B = 2
6. A standardized u chart is being used. Calculate the plot point value using the following information.

   Number of defects per unit found in recent subgroup for part C: 8

   Historical process average for part C = 7

   Historical standard deviation for part C = 1.2
7. Using standardized attribute charts, false signals do not occur on the control chart even when different processes with different process ________________ and different standard deviations are plotted on the same chart.
8. Control limits on standardized short-run attributes charts never need to be recalculated. (True or False)
9. Using standardized attribute charts, historical values (of the average and standard deviation) for the part need to be updated in cases where the process remains in statistical control. (True or False)
10. Sample sizes taken at each subgroup have no affect on the control limits of a standardized attribute control chart. (True or False)

CHAPTER 6

Pattern Analysis

INTRODUCTION

A wide variety of patterns may be seen on a control chart, and an even wider variety of causes make these patterns appear. All processes have common and special causes of variation that are different. One purpose of control charts is to detect special causes of variation so that they can be eliminated. The number or degree of special causes that exist depends largely on the condition of the process. These special causes can include wear and tear, operator intervention or skill, speeds and feeds, coolant, tooling, methods, and measurements. For example, a new process (with new machine, equipment, tooling, and so on) would be expected to have less variation (a better natural tolerance) than an old process, and it often does.

As the example demonstrates, new processes are expected to be better, and most of the time, this is true. One exception to this could be that the wrong new process was selected to do the job. For example, a new drill press is purchased or selected to produce hole sizes that require the inherent accuracy of a broaching machine. The drill press, even though it is new, would have trouble producing the hole sizes that accurately.

Control limits are established by the inherent variation in a process. These limits, remember, are not related to the specification limits. In the previous example, the new drill press is incapable of producing the holes consistently to specifications even though it is a new process. The drill press process is in control, but produces products that are out of specification limits.

Some patterns are commonly seen on control charts. If one has an understanding of these patterns and how to interpret them, control charts can be a powerful tool in achieving a controlled and capable process. This chapter shows some typical control chart patterns and gives examples of each pattern on various control charts. There are also some ideas on special causes that could have caused these particular patterns to appear.

BASIC CHART INTERPRETATION

When control charts are used, there is an easy method to determine if the process is in statistical control or not. The eight statements in Figure 6.1 can be used to see if any control chart is in control. The statements must be answered *true* or *false* while looking at the control chart in Figure 6.2. If all of the statements are *true,* the process is in control. If any statement is *false,* the process is not in control. Several false answers on one process usually means that the process is in a worsened condition.

The eight basic statements of control are effective for a quick look to determine whether statistical control exists, but the following is a more definite set of rules that can be used for individuals and averages charts. Figure 6.3 represents eight tests for special causes (out-of-control conditions). The figure divides the plus and minus three standard deviations from the mean into zones A, B, and C for ease of interpretation. Notes on these tests for special causes are shown in Figure 6.4.

Statements	Answer	
1. There are no plotted points outside the control limits (points plotted on a limit are included).	T	F
2. The total number of points above the centerline is about the same as the total number of points below the centerline.	T	F
3. The plotted points seem to be randomly falling over and under the centerline.	T	F
4. There are no consecutive runs of seven or more points on one side of the centerline.	T	F
5. There are no upward or downward trends of six or more points heading directly toward either control limit.	T	F
6. Only a few of all the points are near the control limits.	T	F
7. The plotted points do not appear to be hugging the centerline with the little distance between them.	T	F
8. There are no straight line patterns.	T	F

Figure 6.1. Eight basic statements of control.

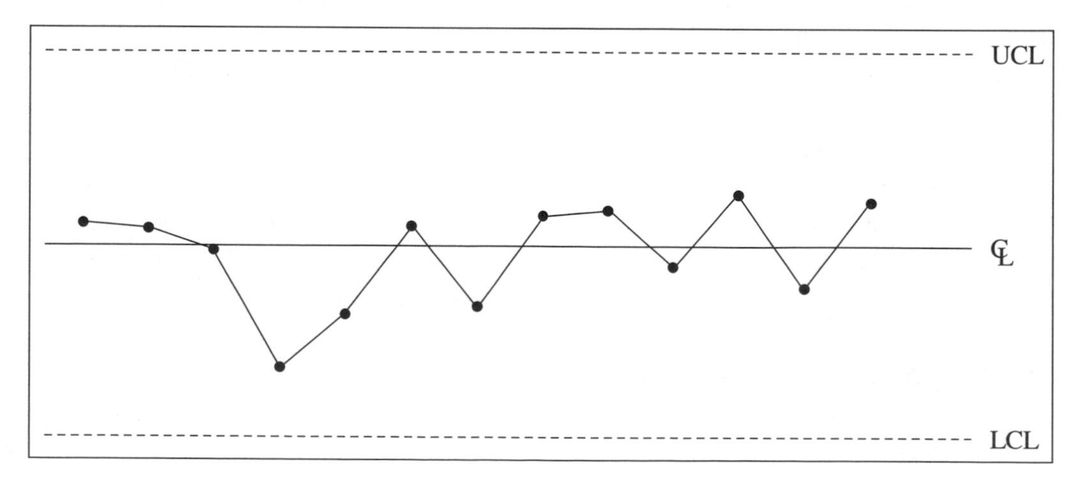

Figure 6.2. A process in control.

Test 1. One point beyond zone A

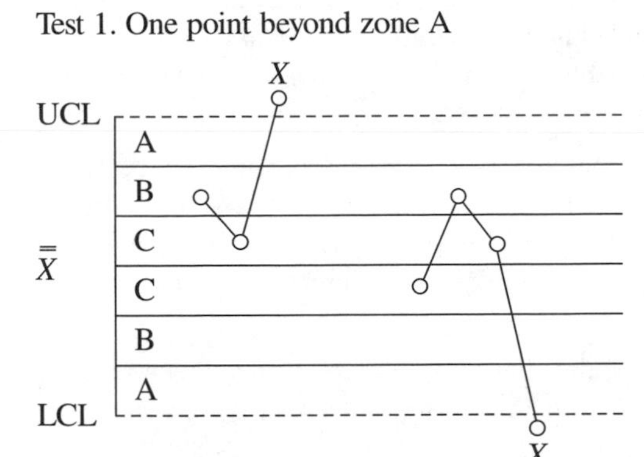

Test 2. Nine points in a row in zone C or beyond

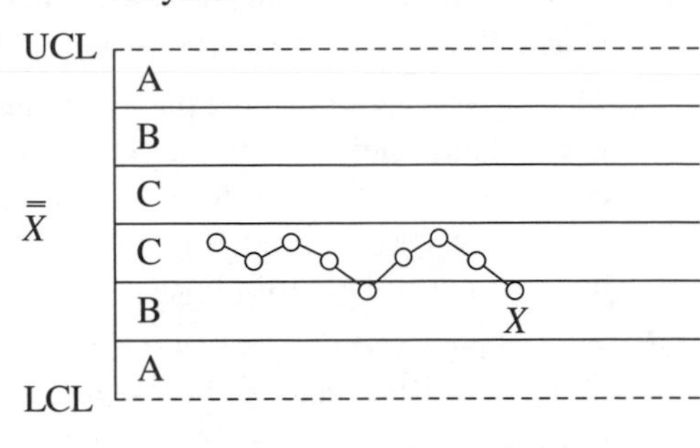

Test 3. Six points in a row steadily increasing or decreasing

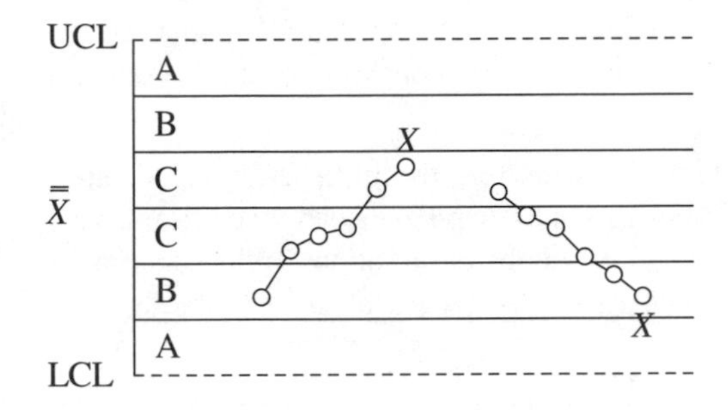

Test 4. Fourteen points in a row alternating up and down

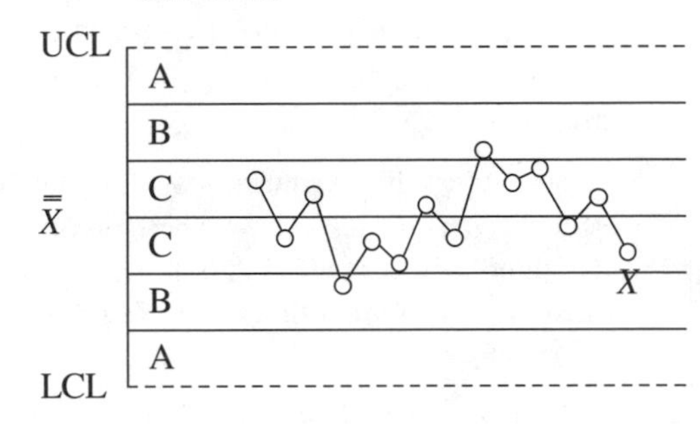

Test 5. Two out of three points in a row in zone A or beyond

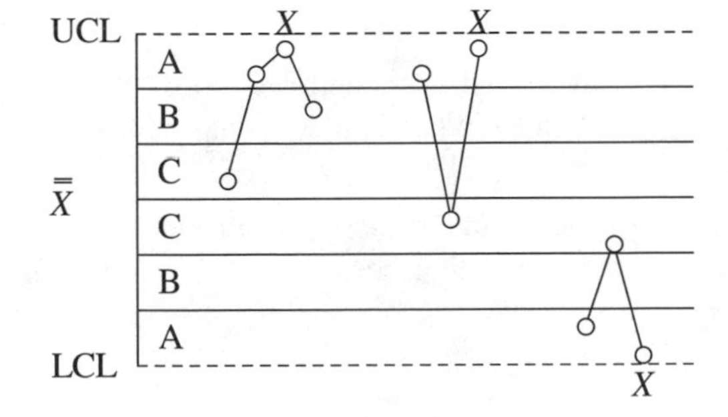

Test 6. Four out of five points in a row in zone B or beyond

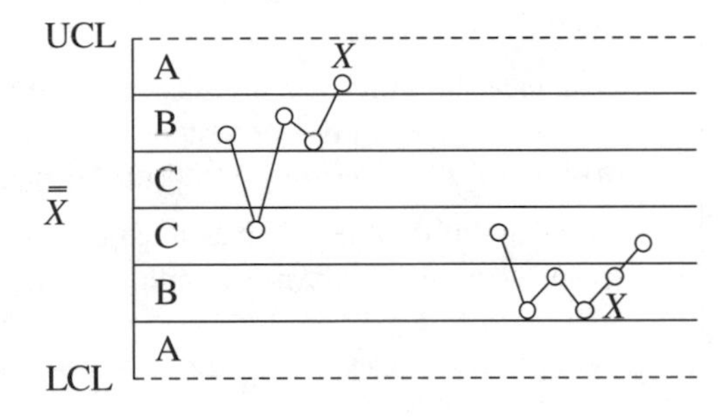

Test 7. Fifteen points in a row in zone C (above and below centerline)

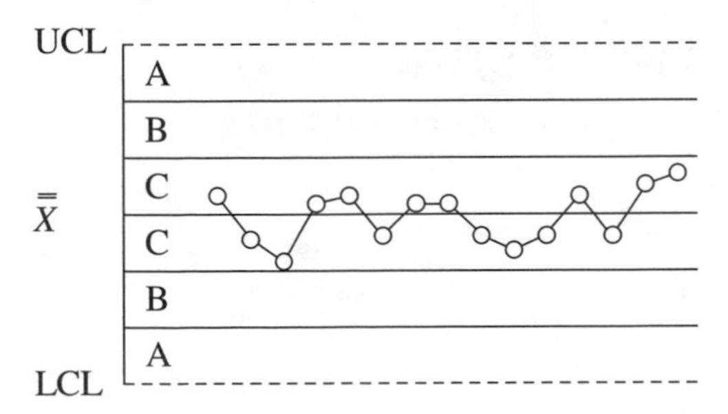

Test 8. Eight points in a row on both sides of centerline with none in zones C

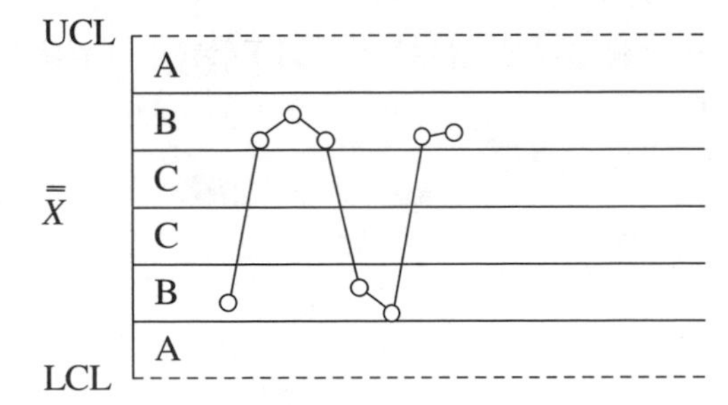

Nelson's tests; courtesy of ASQC *Journal of Quality Technology*

Figure 6.3. Tests for special causes.

1. These tests are applicable to $\overline{X}$ charts and to individuals (X) charts. A normal distribution is assumed. Tests 1, 2, 5, and 6 are to be applied to the upper and lower halves of the chart separately. Tests 3, 4, 7, and 8 are to be applied to the whole chart.
2. The upper control limit and the lower control limit are set at three sigma above the centerline and three sigma below the centerline. For the purpose of applying the tests, the control chart is equally divided into six zones, each zone being one sigma wide. The upper half of the chart is referred to as A (outer third), B (middle third), and C (inner third). The lower half is taken as the mirror image.
3. When a process is in a state of statistical control, the chance of (incorrectly) getting a signal for the presence of a special cause is less than five in a thousand for each of these tests.
4. It is suggested that tests 1, 2, 3, and 4 be applied routinely by the person plotting the chart. The overall probability of getting a false signal from one or more of these is about one in a hundred.
5. It is suggested that the first four tests be augmented by tests 5 and 6 when it becomes economically desirable to have earlier warning. This will raise the probability of a false signal to about two in a hundred.
6. Tests 7 and 8 are diagnostic tests for stratification. They are very useful in setting up a control chart. These tests show when the observations in a subgroup have been taken from two (or more) sources with different means. Test 7 reacts when the observations in the subgroup always come from both sources. Test 8 reacts when the subgroups are taken from one source at a time.
7. Whenever the existence of a special cause is signaled by a test, this should be indicated by placing a cross just above the last point if that point lies above the centerline, or just below it if it lies below the centerline.
8. Points can contribute to more than one test, however, no point is ever marked with more than one cross.
9. The presence of a cross indicates that the process is not in statistical control. It means that the point is the last one of a sequence of points (a single point in test 1) that is very unlikely to occur if the process is in statistical control.
10. Although this can be taken as a basic set of tests, analysts should be alert to any patterns of points that might indicate the influences of special causes in their process.

Nelson's tests; courtesy of ASQC *Journal of Quality Technology*

Figure 6.4. Notes on tests for special causes.

AVERAGES CHART PATTERNS AND POSSIBLE CAUSES

Figures 6.5–6.9 show some examples of averages chart conditions and list some possible causes.

Condition/description

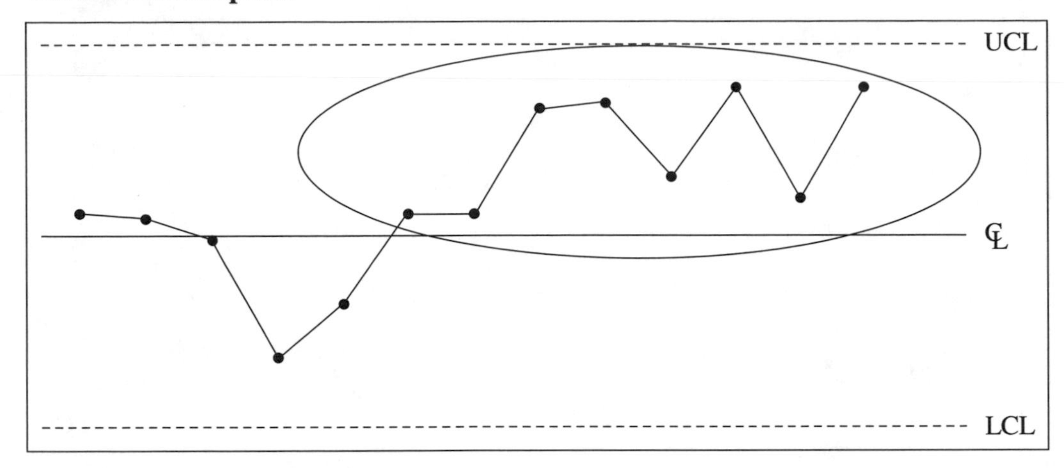

Figure 6.5. Run (jump shift) pattern.

Possible causes

1. Change in machine setting
2. Different operator
3. Different material, method, process
4. Minor failure of a machine part
5. Measuring equipment setting/technique
6. Adjustment on individuals (overcontrol)
7. New gage
8. Fixture change
9. Parts changed

Condition/description

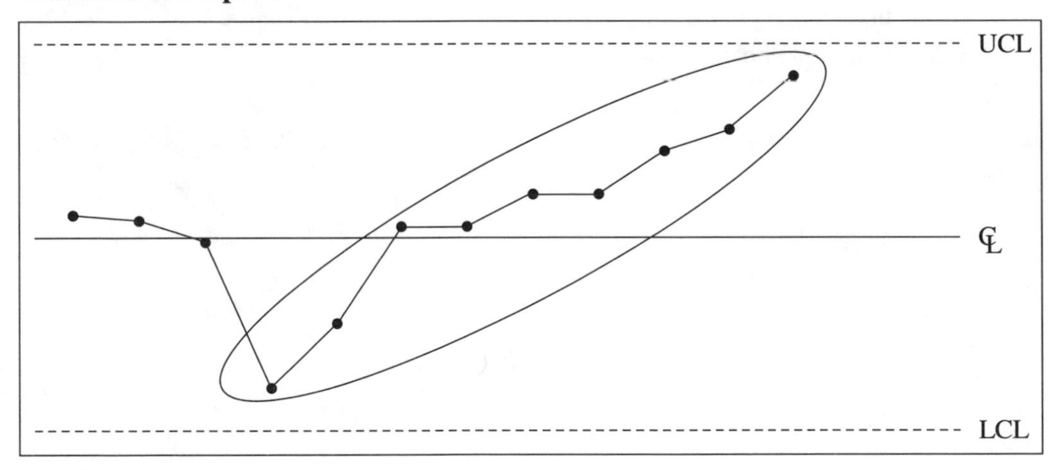

Figure 6.6. Upward trend pattern.

Possible causes

1. Tool wear
2. Gradual equipment wear
3. Seasonal effects (temperature/humidity)
4. Dirt/chip buildup on work-holding devices
5. Operator fatigue
6. Change in coolant temperature

Condition/description

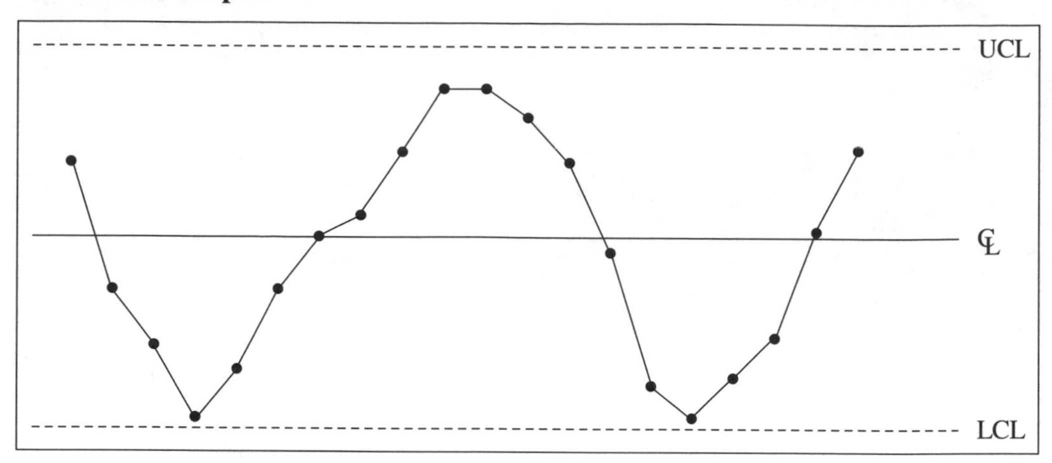

Figure 6.7. Recurring cycles pattern.

Possible causes

1. Different incoming materials
2. Cold startup
3. Seasonal effects (temperature/humidity)
4. Voltage fluctuations
5. Merging of different processes
6. Chemical or mechanical properties
7. Periodic rotation of operators
8. Measuring equipment not precise
9. Calculation and plotting mistakes
10. Machine won't hold setting

Condition/description

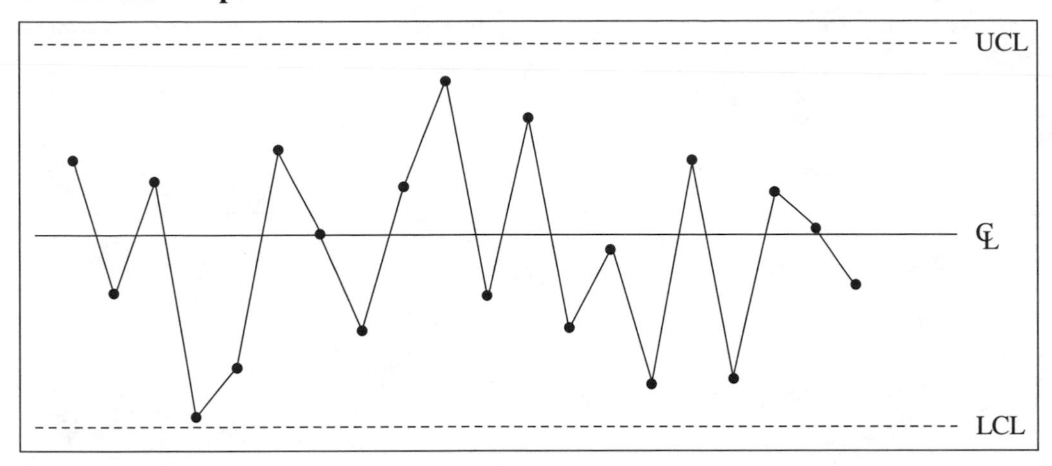

Figure 6.8. Two universe pattern.

Possible causes

1. Large differences in material quality
2. Two or more machines using the same chart
3. Within the piece, variation not considered (such as taper/roundness)
4. Large differences in the method of measurement of the product

Condition/description

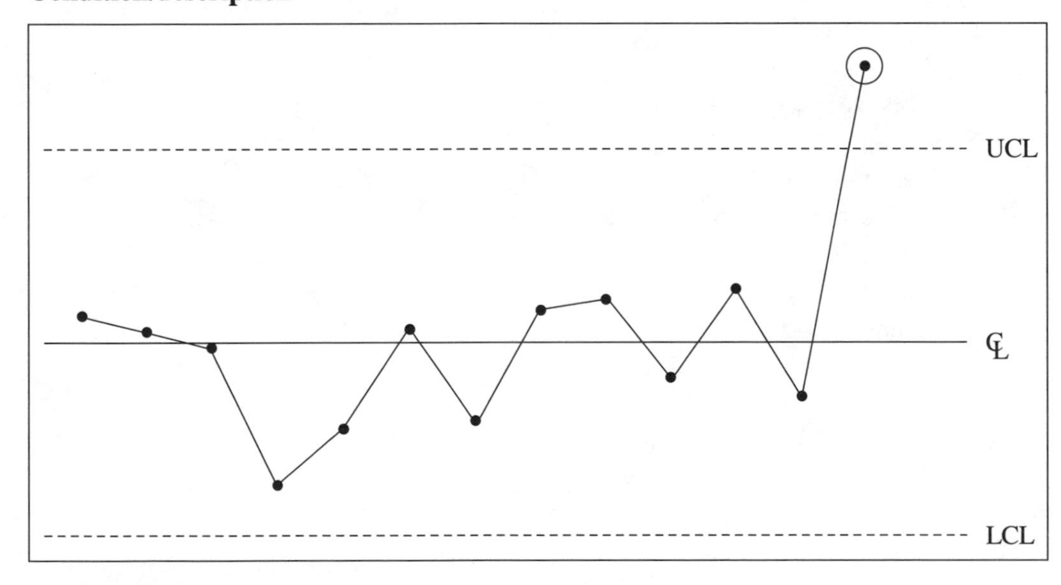

Figure 6.9. Freak (one point out).

Possible causes

1. Power surge
2. Hardness of a single part
3. Broken tool
4. Gage jumped setting

RANGE CHART PATTERNS AND POSSIBLE CAUSES

Figures 6.10–6.15 show some range chart conditions and list possible causes for those conditions.

Condition/description

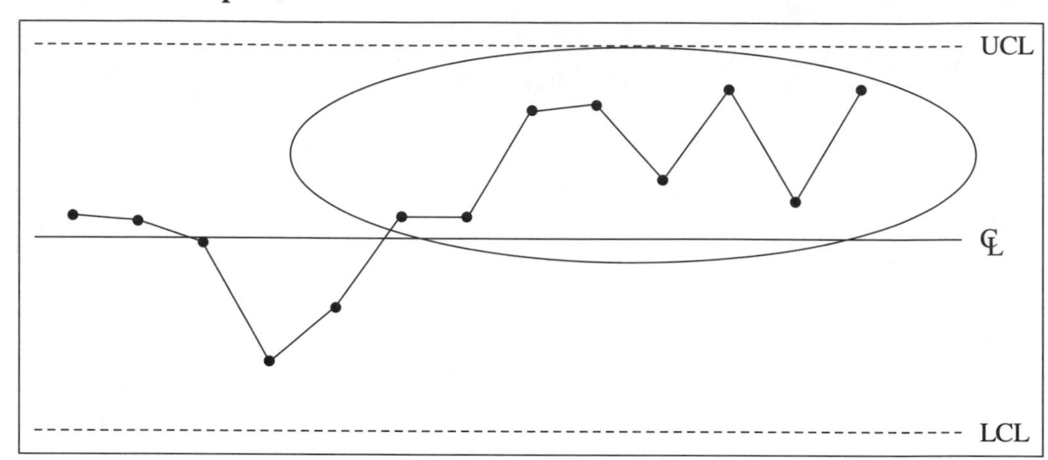

Figure 6.10. Run (jump shift) pattern.

Possible causes

1. Sudden increase in gear play
2. Greater variation in incoming material
3. Inexperienced operator
4. Miscalculation of ranges
5. Excessive speeds and feeds
6. New operator
7. Change in methods
8. Long-term increase in process variability
9. Gage drift
10. New tools
11. Fixture change

Condition/description

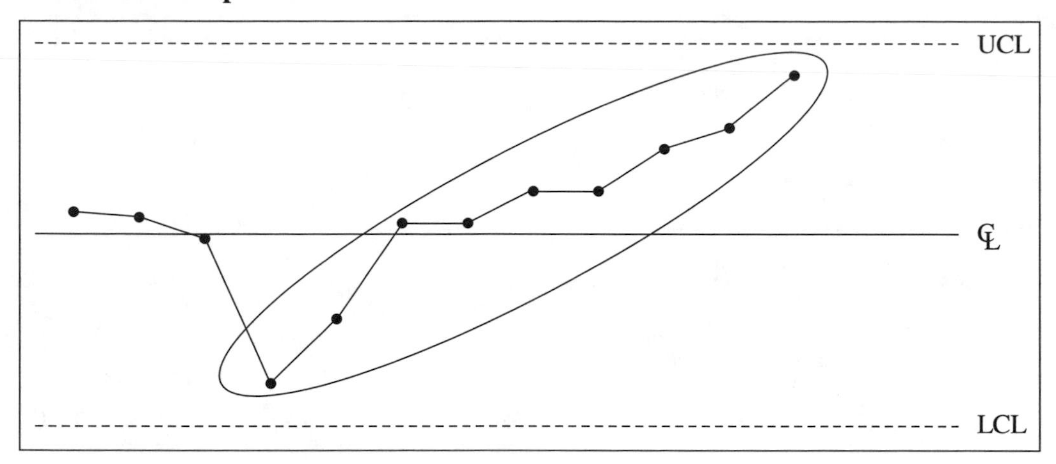

Figure 6.11. Upward trend pattern.

Possible causes

1. Decrease in operator skill due to fatigue
2. A gradual decline in the homogeneity of incoming material
3. Some machine part/fixture loosening
4. Gage drift
5. Deterioration of maintenance
6. Tool wear

Condition/description

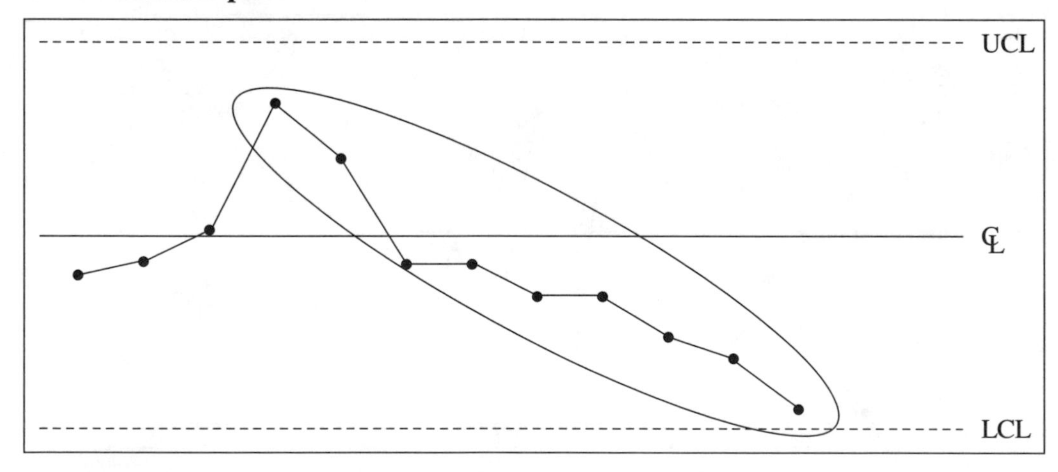

Figure 6.12. Downward trend pattern.

Possible causes

1. Improved operator skill
2. A gradual improvement in the homogeneity/uniformity of incoming material
3. Better maintenance intervals/program
4. Previous operation is more uniform in its output

Condition/description

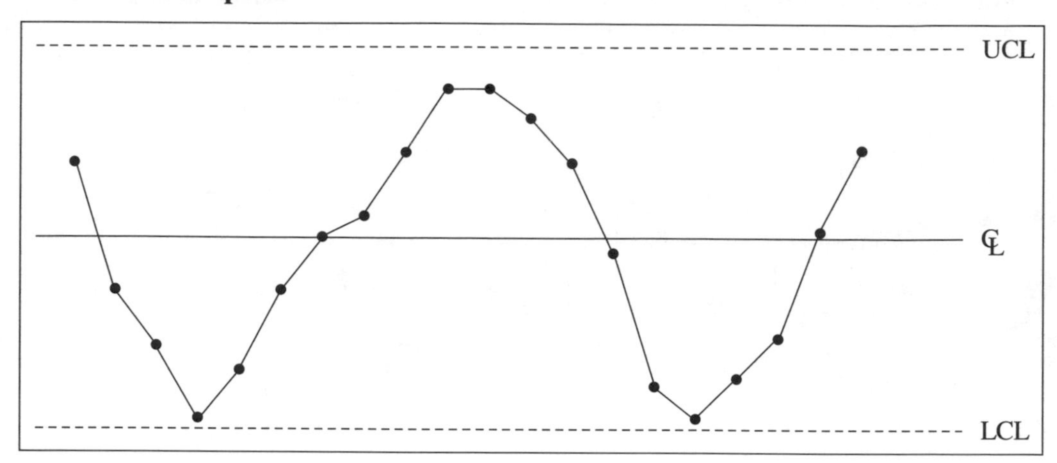

Figure 6.13. Recurring cycles pattern.

Possible causes

1. Operator fatigue and rejuvenation due to periodic breaks
2. Lubrication cycles
3. Rotation of operators, fixtures, gages
4. Differences between shifts
5. Worn tools
6. Differences between machine needs

Condition/description

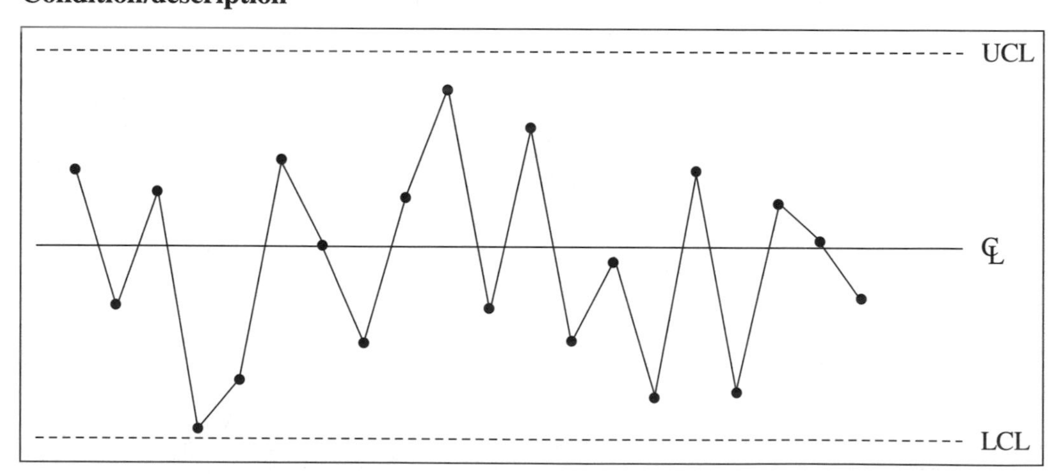

Figure 6.14. Two universe pattern.

Possible causes

1. Different machines or operators using the same chart
2. Materials used from different suppliers

Condition/description

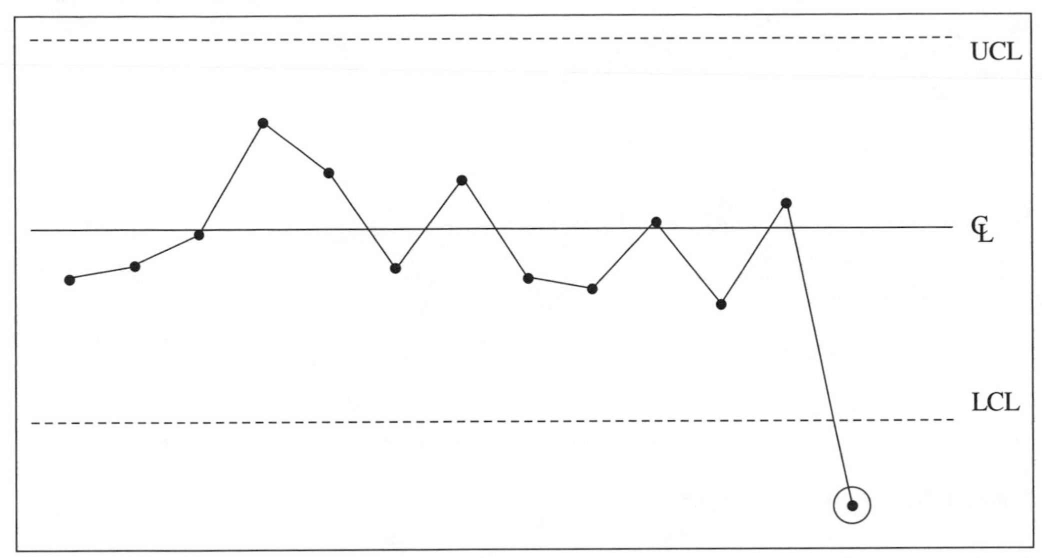

Figure 6.15. Freak (one point out).

Possible causes

1. Power surge
2. Hardness of a single part
3. Tool broke
4. Gage jumped setting

INTERPRETING AVERAGES AND RANGE CHARTS TOGETHER

Averages and range charts must be interpreted together as well as separately. A stable process will have points distributed between control limits on the charts randomly. If the process is stable, the points on the averages chart and the ranges chart should not tend to follow each other. Lack of stability will sometimes cause the two charts to move together.

For example, if a process is *positively* skewed, then the points tend to correlate positively with each other (points are high on both charts). If a process is *negatively* skewed, then the points tend not to correlate (points on the averages chart will follow points on the ranges chart but in opposite directions).

Skewness is a measure of the symmetry of the distribution (refer to appropriate statistical textbooks for detailed coverage on skewness). Positive skewness is when the distribution slopes downward to the right and negative skewness is when the distribution slopes downward to the left.

p CHART PATTERNS AND POSSIBLE CAUSES

Figures 6.16–6.21 show *p* chart patterns and list possible causes.

Condition/description

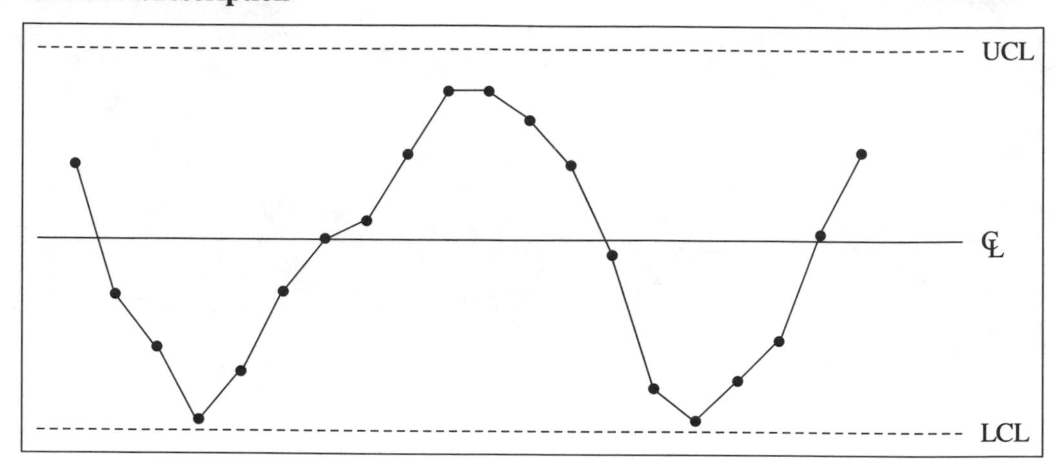

Figure 6.16. Recurring cycles pattern.

Possible causes

1. Considerable sorting
2. Sampling practices
3. Different suppliers
4. Different processes

Condition/description

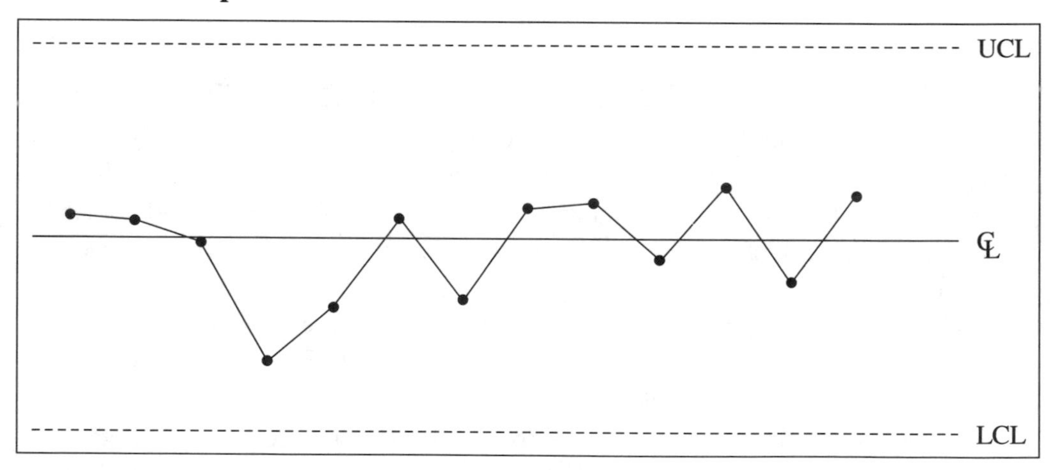

Figure 6.17. Natural pattern, which indicates that the fraction defective is constant in the process and that sampling is at random.

Condition/description

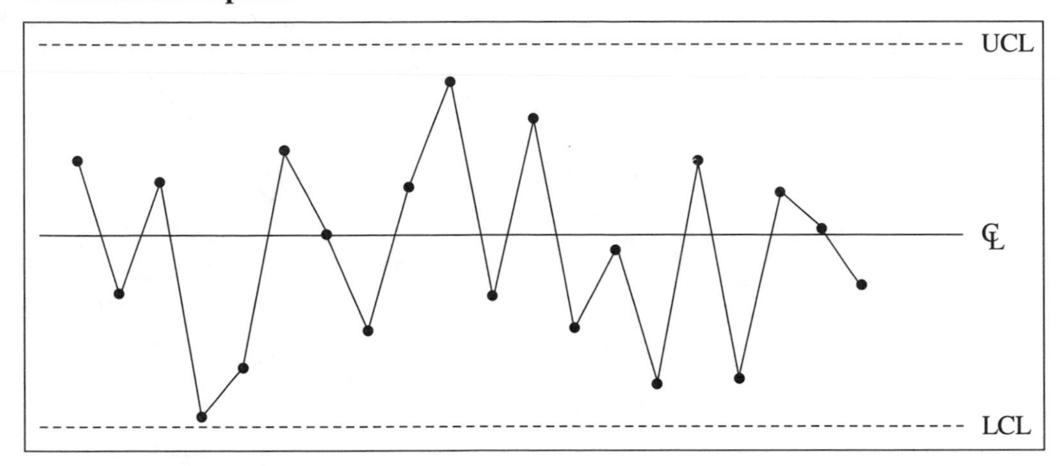

Figure 6.18. Two universe pattern.

Possible causes

1. Nonrandom sampling
2. Differences in gages or measurements
3. Different sources
4. Previous operations are sorting lots
5. Different standards
6. Different inspectors

Condition/description

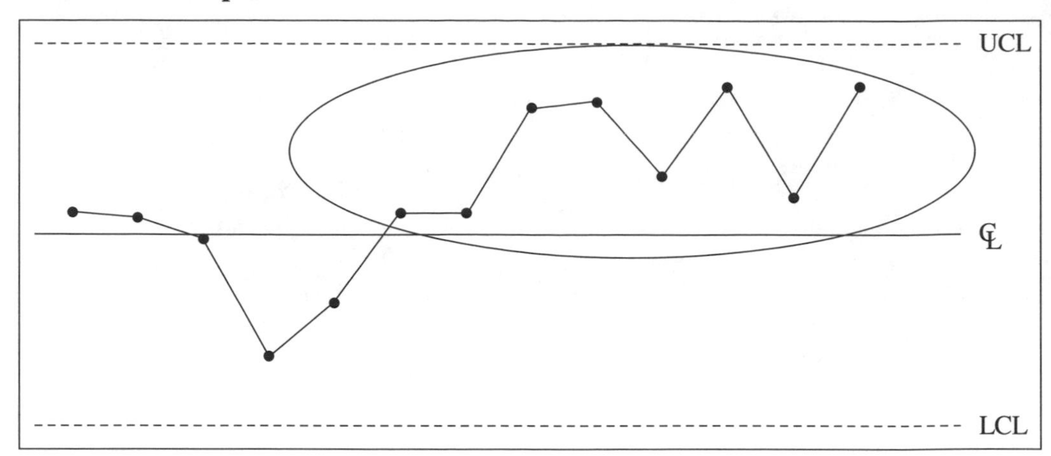

Figure 6.19. Run (jump shift) pattern.

Possible causes

1. New materials
2. Changed standards
3. Changed operators
4. Changed methods
5. Measurements have changed

Higher level:

1. Materials are worse
2. Machines, tooling, and so on worsening
3. New operator
4. Tightened standards

Lower level:

1. Better operators
2. Better machines, tooling, and so forth
3. Better methods or materials
4. Loosened standards

Condition/description

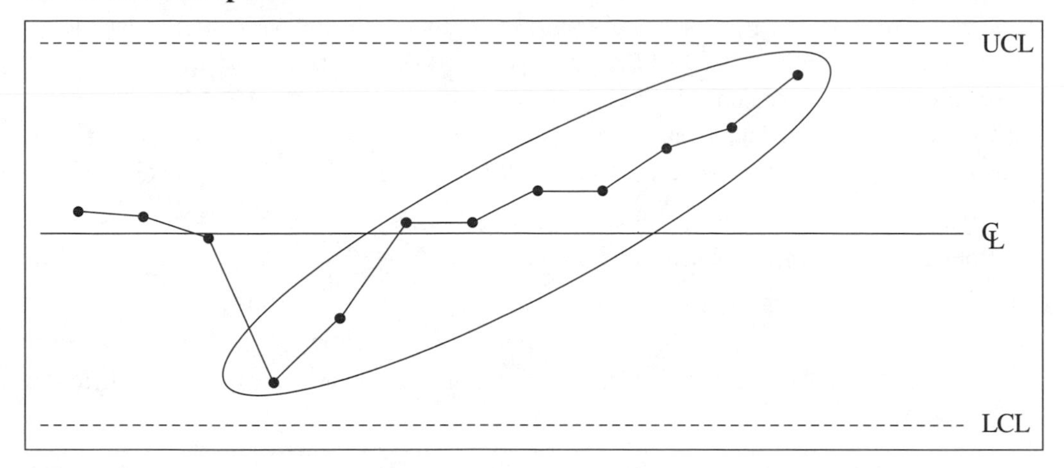

Figure 6.20. Upward trend pattern.

Possible causes

1. Poorer materials
2. Tool wear
3. Gage drifts
4. Tightened standards
5. Poorer work by the operator

Condition/description

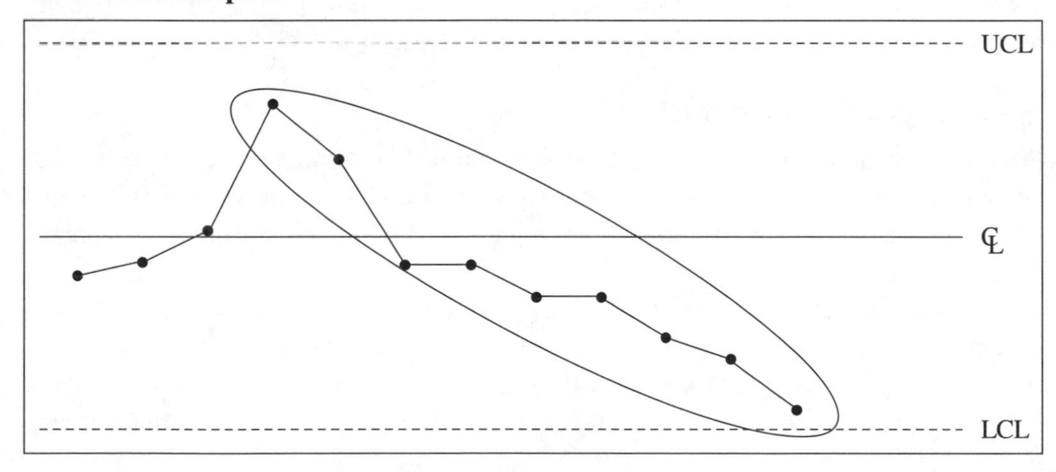

Figure 6.21. Downward trend pattern.

Possible causes

1. Better skill of the operator
2. Better materials or tools
3. Loosened standards
4. Better methods

OTHER STATISTICAL CONTROL TESTS

Another popular set of tests for out-of-control conditions are the Western Electric Tests. These tests, just like Nelson's tests previously covered, assist the processor in determining if the process is in or out of control based on the patterns of plot points. The Western Electric rules are somewhat less definitive (and there are fewer of them) than the Nelson's tests. The following is a review of the Western Electric rules and an example of the unstable patterns they represent.

Rule 1. Any one plot point beyond a three-sigma limit. This rule is looking for freak points.

Rule 2. Two out of three successive plot points beyond either of the two-sigma limits from the mean. This rule is looking for the beginning of runs, cycles, or trends in the process.

Rule 3. Four out of five successive plot points beyond either of the one-sigma limits from the mean. This rule is looking for runs, cycles, or trends in the process.

Rule 4. Eight successive plot points on the same side of the centerline. This rule is looking for runs in the process.

PATTERN ANALYSIS

Short-Run Target Charts

For all short-run target charts, interpretation of chart patterns does not change. The same types of patterns (runs, trends, and so on) will be seen on a short-run target chart as they might be seen on a traditional control chart. The primary difference, for variables charts, is the range test that watches for high or low run patterns on the range chart immediately after a new part number has been set up and plotted. The major benefit is that every time the process goes out of control, and the assignable cause is removed, the reduced variability affects all part numbers that will be made on the process.

Short-Run Standardized Charts

Standardized charts are very different when it comes to pattern analysis. Since each and every plot point is standardized and could represent a different part number or different process altogether, specific patterns do not represent a continuous stream of variation as with traditional charts.

One method that may be helpful is to vary the styles of the plot points for each different part/process. For example, small circles could be used for plot points of process A, and solid dots could be used to plot process B. In this manner, one could watch for trends in process A or B by interpreting (visually connecting) only the points from that process. Refer to this discussion in chapter 5, specifically in Figure 5.9.

For most short-run process control applications, I prefer to use the different types of target charts instead of the standardized charts because standardized charts can be very confusing to the operators. There is a lot to be said for keeping it as simple as possible in the application.

OUT-OF-CONTROL CONDITIONS

The primary purpose of a process control chart is to identify when special causes exist in the process so that real-time action can be taken to eliminate the cause(s) and get the process in control. In order for process control to work then, users must be prepared to take action when the process goes out of control. Successful reaction to out-of-control conditions in the process is a matter of interpreting control charts for special causes and keeping in mind which chart shows the out-of-control condition and what that chart represents. Charts for averages, for example, control the central tendency (or setup) of the process, and charts for ranges control the variability. It is useless to use statistical control charts unless the user is prepared to correct the process when it goes out of control.

It is a good idea, during process control planning, to evaluate the process for the purpose of identifying specific sources of special causes. Methods for evaluating process for this purpose include brainstorming, cause-and-effect diagrams, design of experiments, or process failure modes and effects analysis (PFMEA). For example, a failure modes and effects analysis (and brainstorming) for a lathe that turns diameters and faces could result in possible sources of special causes for the average and the range chart. Tool wear could be classified as a possible special cause for trends in the averages and/or the range chart, and loose fixture could be the cause for a run pattern in the averages chart. Material hardness may fall in the category that affects the ranges more than the averages, and so on. The most frequently used approach is to use the control charts and take notes on actions taken versus the special causes that they correct. Whenever a process goes out of control and an action is taken that brings the process into control, there is strong support for the hypothesis that the special cause has been found and corrected. If an action that is taken does not bring the process back into control, it is a good bet that the cause has not yet been found. Refer to the charts and possible causes covered previously in this chapter. These causes are actual cases where the pattern was seen on an actual control chart and the action that was taken brought the process into control.

Variables Charts

In general, if the averages chart shows out-of-control patterns and the range chart is in control, the special causes that exist are affecting the setup (or central tendency) of the process. In these cases, the user should look for problems affecting the setup such as tool wear, overadjustment, loose fixtures, and many other problems that affect the average size or level of the process, but not necessarily the part-to-part (or time-to-time) variation.

If the chart for ranges goes out of control and the averages chart is in control, the user should look for special causes that affect part-to-part (or time-to-time) variation, but do not affect the setup. If both average and range chart go out of control, then the special cause is one that affects the setup of the process and the part-to-part (or time-to-time) variation.

Attribute Charts

Attribute charts are those that monitor the fraction (or proportion) defective (p), number of defectives (np), number of defects (c), or the number of defects per unit (u). In most cases, attribute charts have a lower control limit of zero (when the lower control limit calculation results in a negative number). When an attribute chart shows out-of-control conditions, one must consider what the chart is monitoring and what the pattern means to the process. For example, an out-of-control pattern such as a downward trend on any of these charts means that there are special causes in the process that are resulting in gradual improvement. An upward trend on these charts means that there are special causes in the process that are resulting in the gradual degradation of quality levels.

In all cases, out-of-control conditions in the process means that something has changed (for the worse or for the better), but the special cause needs to be identified in order to correct the cause (if it means a worsening affect) or implement the cause (if it means improvement). Other patterns of out-of-control conditions mean that special causes exist and the process average ($\overline{p}$, $n\overline{p}$, $\overline{c}$, $\overline{u}$) is unstable.

REVIEW PROBLEMS

Refer to Appendix H for answers.

Using the eight basic statements of control, look at the following five charts and determine if they are in statistical control. Answer *yes* or *no*.

1.

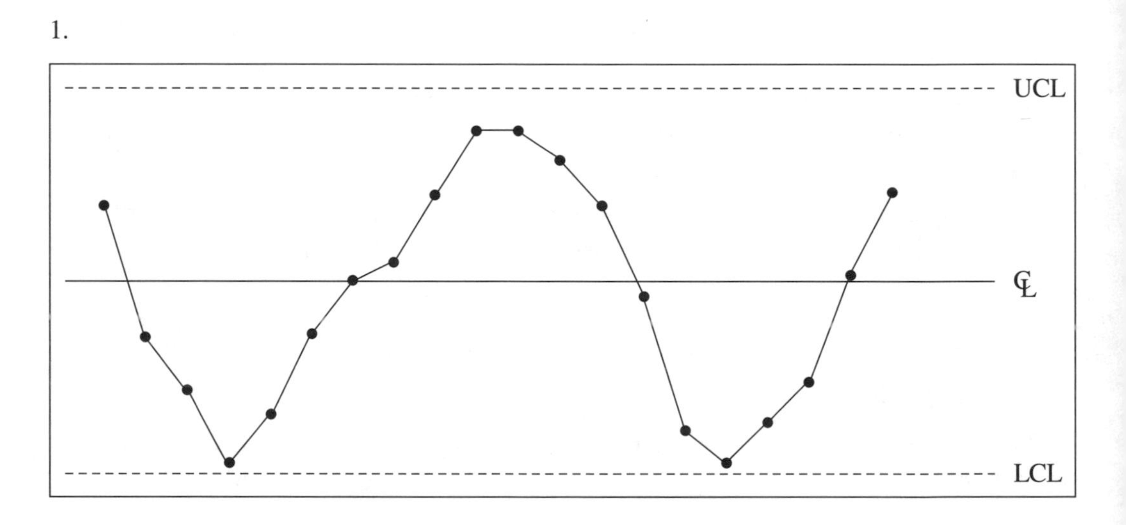

Is this process in control?

2.

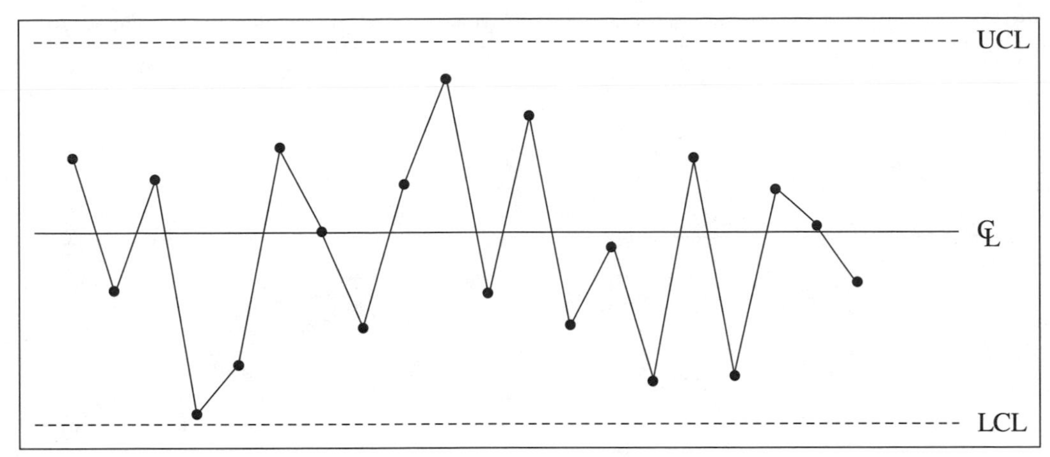

Is this process in control?

3.

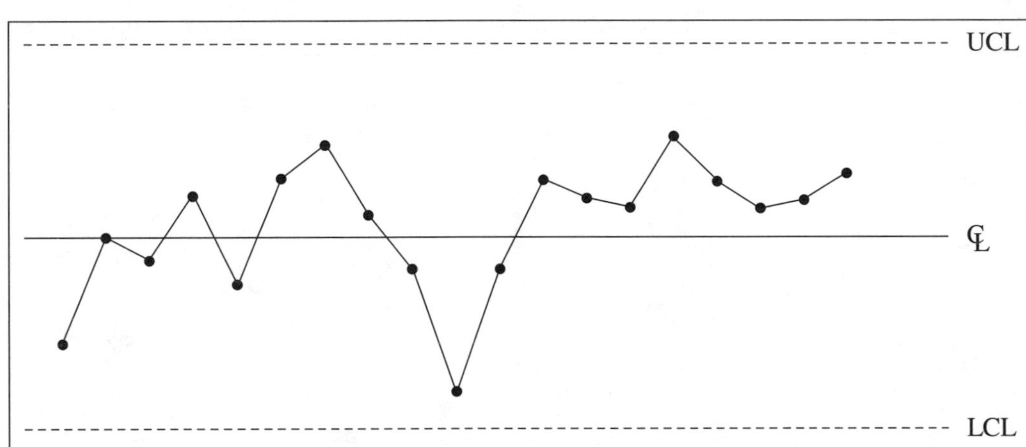

Is this process in control?

4.

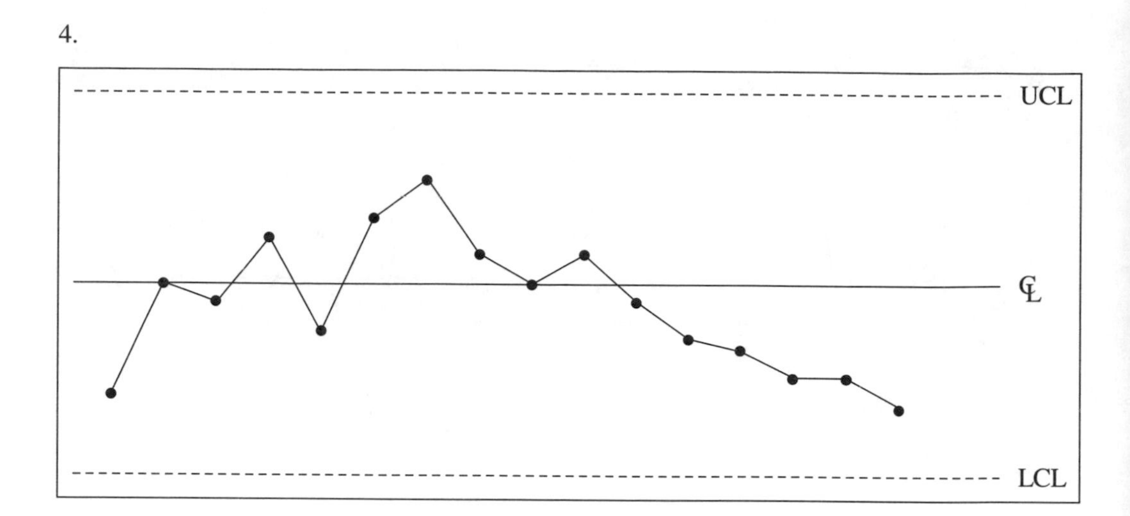

Is this process in control?

5.

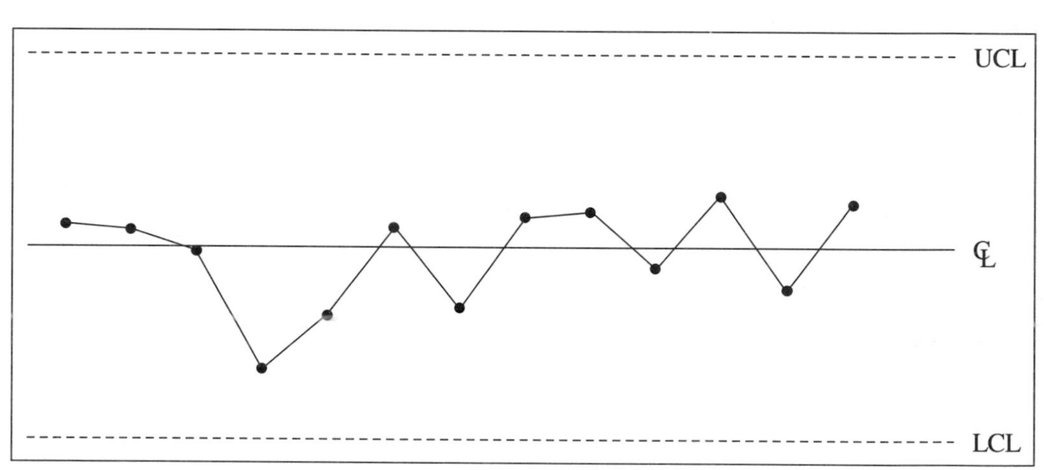

Is this process in control?

CHAPTER 7

Process Capability

INTRODUCTION

When a process is in statistical control (all special causes of variation have been eliminated) and the data distribute normally, the process capability can be studied. (Note: There are a variety of tests (not covered in this book) that can be performed to see if the data are normally distributed. Refer to the bibliography for references on tests for normality.) Also, the process variation consumes no more than a certain percentage of the total specification tolerance. This percentage depends on the company's capability goals (or requirements).

FREQUENCY DISTRIBUTIONS AND HISTOGRAMS

A frequency distribution (see Figure 7.1) is a graph in which the horizontal axis is the scale of the measurements and the vertical axis is the frequency (the number of times a given measurement occurs). The frequency distribution then uses tally marks (or *X*s) to display the frequency of each measurement. If a line is drawn around the plotted tally marks (or *X*s), the shape of the distribution can be seen.

A histogram (see Figure 7.2) is similar to a frequency distribution except bars are used instead of tally marks or *X*s to show the frequency of occurrence of each measurement. A vertical axis scale is also used to show the frequency values.

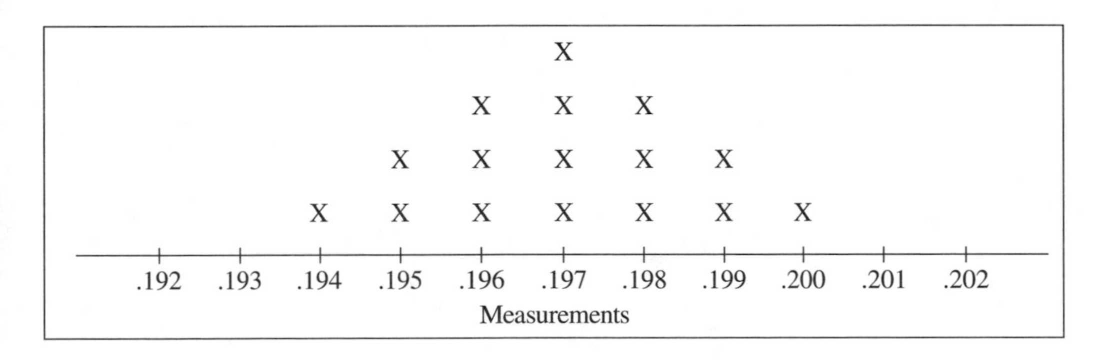

Figure 7.1. Frequency distribution of the data.

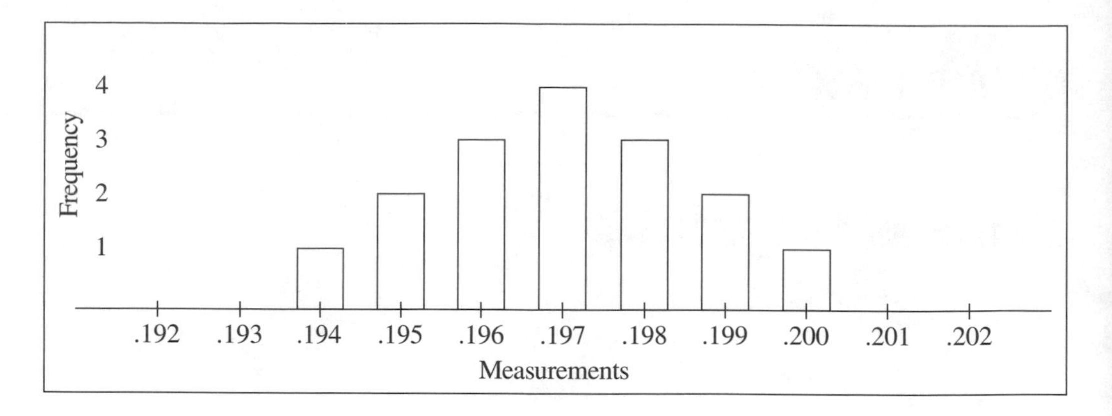

Figure 7.2. Histogram of the data.

THREE PRIMARY CAUSES FOR DEFECTIVE PRODUCTS

There are three primary conditions (or causes) for a process to make defective products. Once these conditions have been corrected, an acceptable yield of good product will result. The three conditions that cause defective product follow.

1. *The process is out of control* (see Figure 7.3). When assignable (or special) causes exist in a process, the process will vary abnormally. Abnormal variation, depending on the capability of the process, can be the cause of defective products.
2. *The process is not centered on specifications* (see Figure 7.4). Processes must be centered on specification tolerances to the degree that the variation of the process will not cause defective products. If a process is significantly off center, the variation

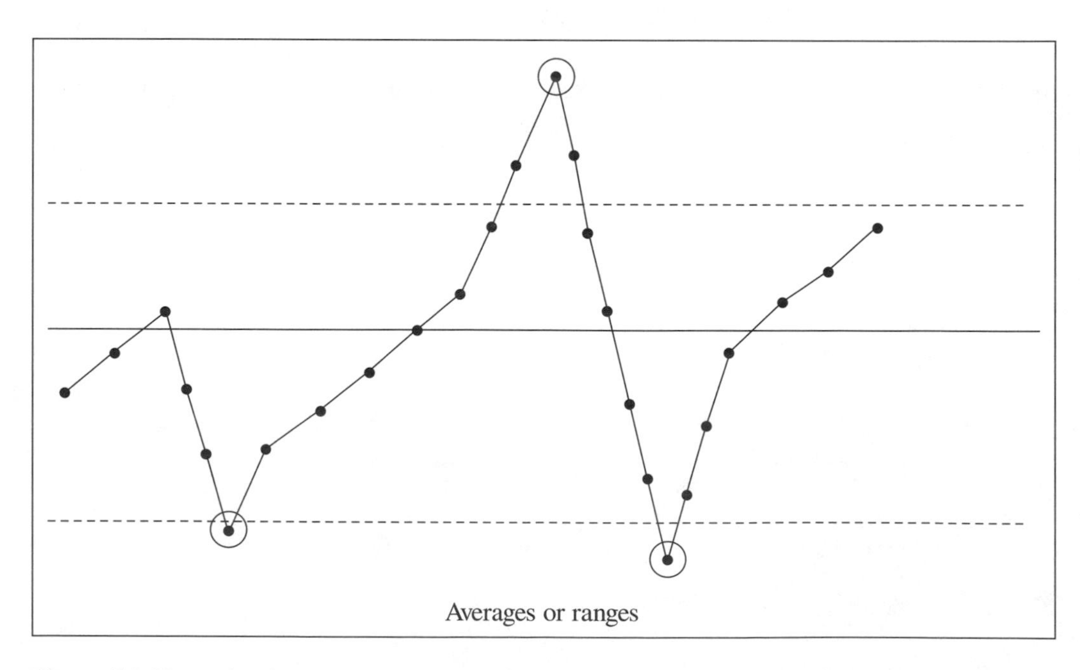

Figure 7.3. Example of a process that is out of control.

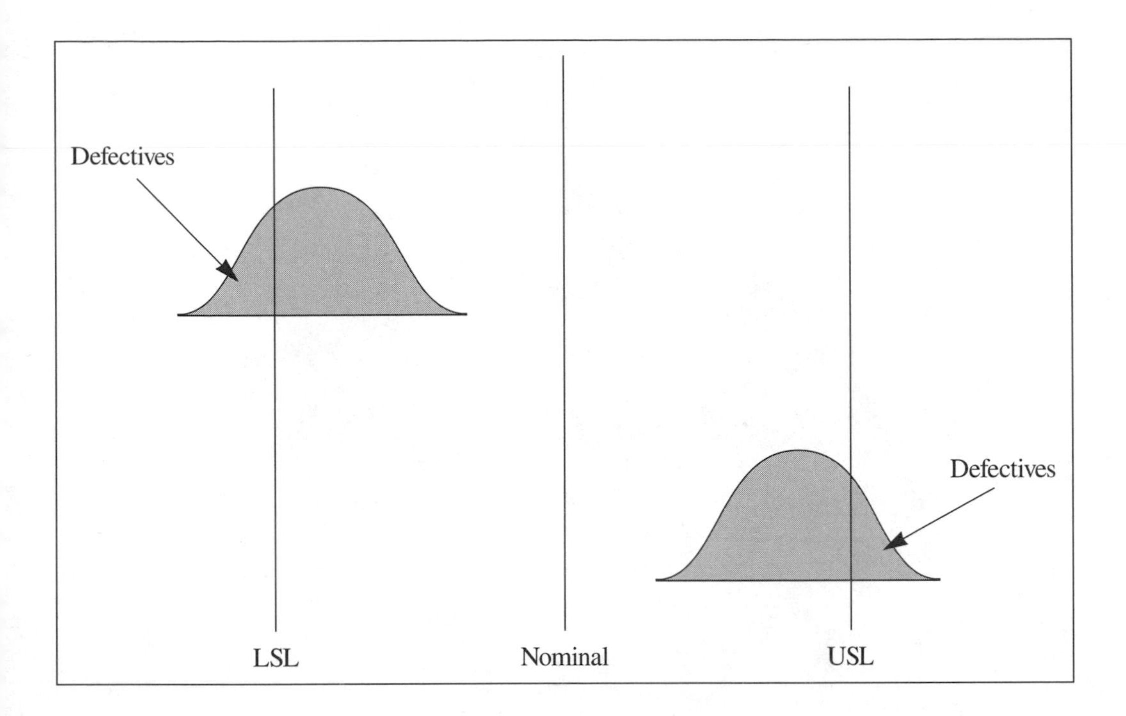

Figure 7.4. Example of a process that is significantly off center.

of the process can exceed either specification limit. From other points of view (such as Taguchi's loss function), processes should be centered as close to nominal as possible regardless of their capability. Refer to other texts that cover Taguchi methods.

3. *The process is not capable* (see Figure 7.5). If a process is not capable (for example, the inherent variation of the process is wider than the total specification tolerance), it will always produce defective products. Centering an incapable process helps to improve the yield, but there will still be defectives.

Statistical process control methods (and teamwork) are the only way to correct these three adverse conditions in a process. The assumption, of course, is that the appropriate statistical methods will be used to control the process, and action will be taken to improve the process condition that is causing defective product. Statistical methods help to identify adverse conditions, understand them, and provide the opportunity to correct them. The statistical method itself (just doing charts) will not correct the process.

THE NORMAL DISTRIBUTION

The normal distribution (see Figure 7.6) is a symmetrical bell-shaped curve. The average ($\bar{X}$) is the center of the curve, and the variation around the average is indicated by an infinite number of standard deviation (sigma) units in both directions (left and right). For practical purposes, the plus and minus three sigma limits around the mean is used. This represents 99.73 percent of the area under the normal curve. The percentage of observations that can be expected to fall within any area under the normal curve can be predicted. Table 7.1 shows specific areas under the normal curve between specific plus and minus sigma limits.

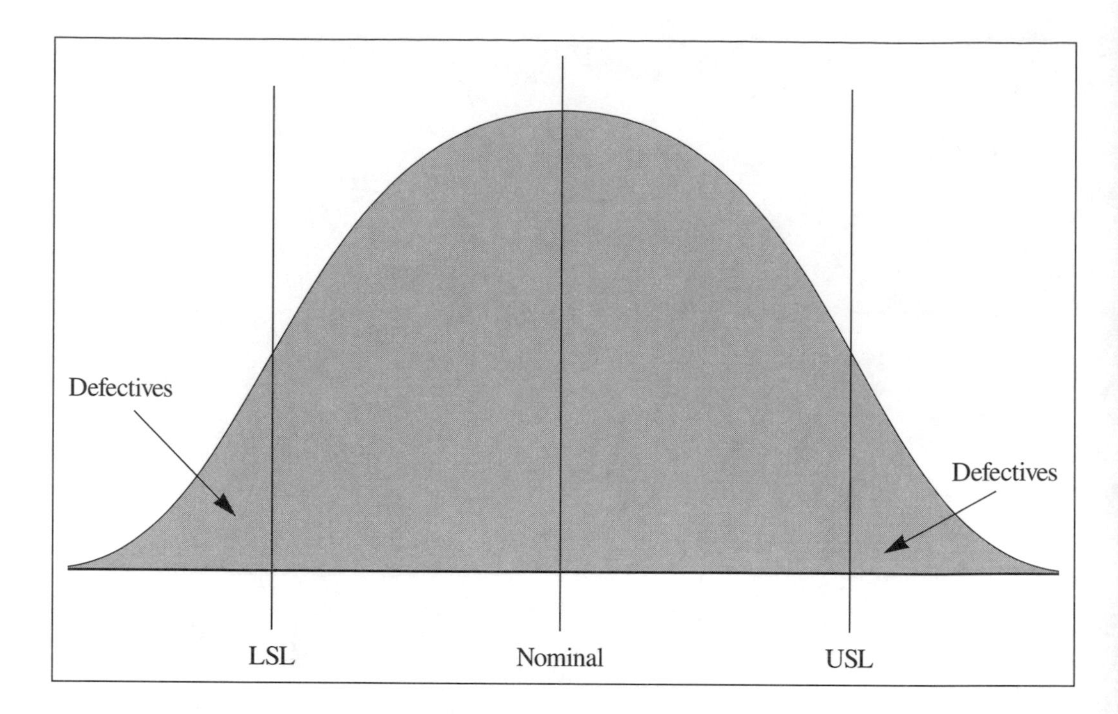

Figure 7.5. Example of a process that is not capable.

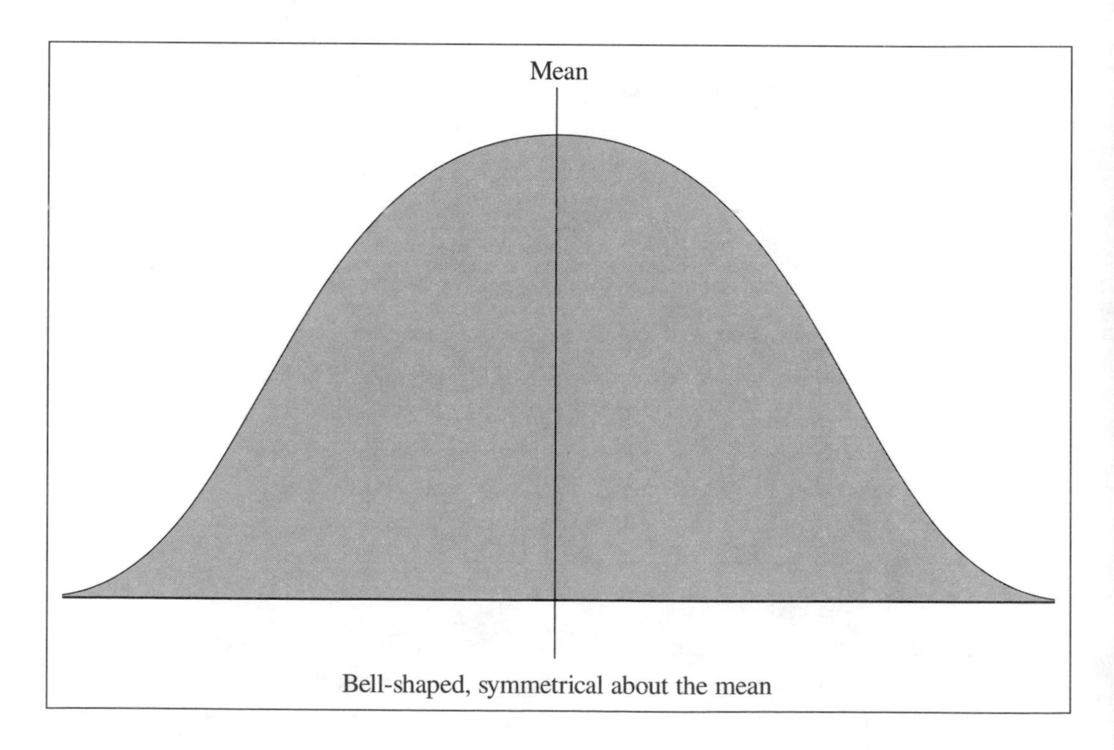

Figure 7.6. The normal distribution or bell curve.

ble 7.1. Area under the curve.

Area under the curve	Percentage
$\bar{X} \pm 1\sigma$	68.26%
$\bar{X} \pm 2\sigma$	95.46%
$\bar{X} \pm 3\sigma$	99.73%
$\bar{X} \pm 4\sigma$	99.994%

Areas Under the Normal Curve

The table of areas under the normal curve (see Appendix A) will be used in later chapters to make predictions about the yield of a process. At this point, it is important to understand how to read the table. The areas under the normal curve between whole standard deviation limits from the mean are shown in Table 7.1. For example, the area between plus and minus one standard deviation from the average represents 68.26 percent of the total area under the curve. The table is used when an area needs to be found using fractional standard deviations from the average (for example, 1.7 standard deviations).

Another example is a normal distribution of data exists from a process that tightens bolts to a specified tolerance of torque.

Specification: 12 ± 3 inch pounds

Process average = 12.5 inch pounds

Standard deviation = 0.7

What percentage of product can be expected to fall between the average (12.5 inch pounds) and 14 inch pounds?

To solve this problem, refer to the table of areas under the normal curve (Appendix A). Calculate the Z value (often referred to as the Z score) and round it off to two decimal places.

$$Z = \frac{|X - \bar{X}|}{s}$$

where

X is a specific limit (for example, specification limit)

$\bar{X}$ is the process average

s is one standard deviation

$$Z = \frac{|X - \bar{X}|}{s} = \frac{|14 - 12.5|}{0.7} = 2.14$$

A Z value of 2.14 means that the specified interval falls between $\bar{X}$ and 2.14 standard deviations.

Find the Z value in the table. The Z value of 2.14 is found by locating 2.1 in the left column and the second decimal place (.04) in the upper row. Converge these two values and find the decimal results, which is .4838. Converted to a percentage, .4838 equals 48.38 percent. This means that 48.38 percent of the product from this process can be expected to fall between 12.5 and 14 inch pounds of torque.

In summary, the table of areas under the normal curve is used to estimate the percentage of the area under the curve that falls between the average ($\bar{X}$) and a specified limit in question. One of the most important things to remember is that, only if the process is in control, can you accurately predict the output of the process.

DISTRIBUTION COMPARISONS

Figure 7.7 shows some comparisons of the center and the variation of processes to specification limits. Processes, even when they are in control, can produce defectives because

1. The process is not centered on target.
2. The variation of the process is too wide.
3. A mixture of center and variation causes some defective parts.

The comparisons in Figure 7.7 illustrate process center and variation problems and the case where the process is both centered and capable.

ALTERNATIVES FOR INCAPABLE PROCESSES

When a process is in control but is not capable of meeting specification limits, there are several alternatives that can be taken. (Note that they are not in priority order.)

1. Repair the process (for example, new bearings, ways, spindles).
2. Modify the process (for example, change the method, speeds, feeds, locators).

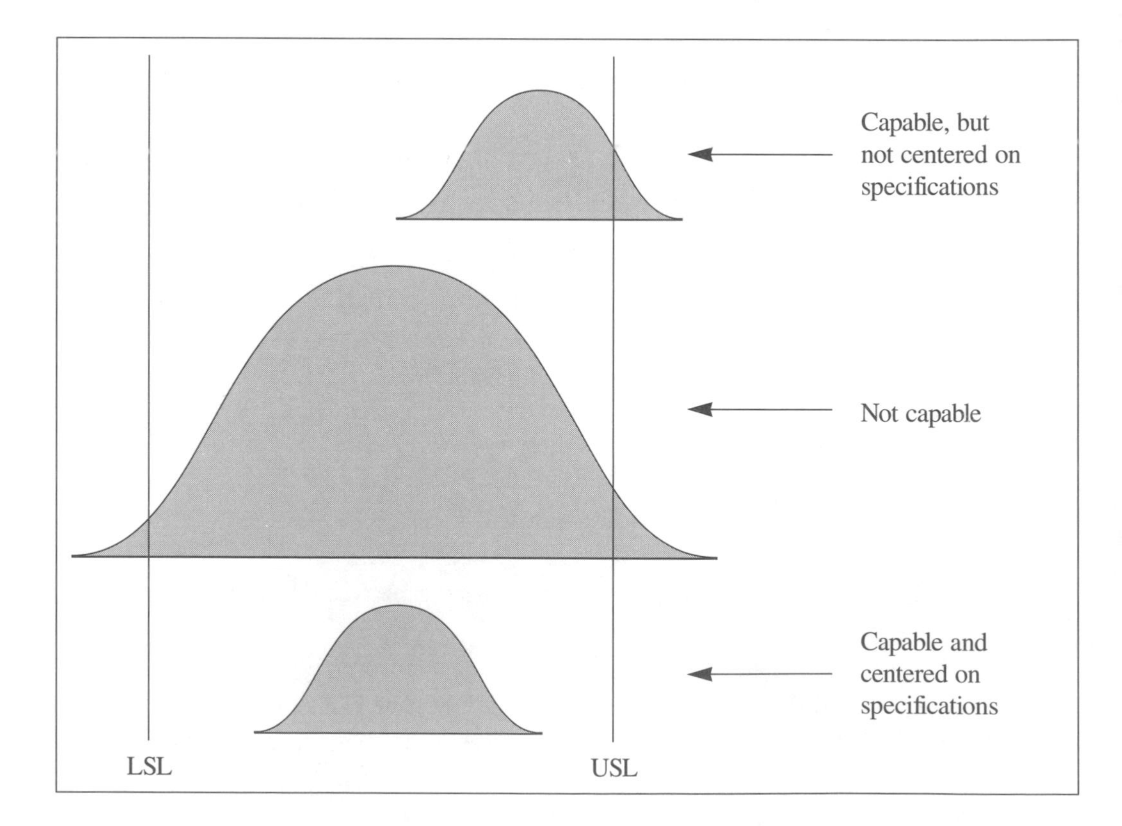

Figure 7.7. Distribution comparisons.

3. Select a better process (for example, grind instead of turn, jig bore instead of drill).
4. Purchase new equipment.
5. Reconsider the make/buy decision. A supplier may be able to produce the product better. Capability should also be studied at the supplier.
6. Inspect the product 100 percent. This decision should be the last resort. Inspection 100 percent is not effective and is costly.
7. Selective assembly or sorting. This alternative is not recommended but applicable in certain special cases. It is, however, time-consuming and costly, and affects interchangeability.
8. Apply problem-solving techniques to improve the capability of the process. See chapter 10 on problem solving.

There are numerous other alternatives that can be considered. The primary goal is to get a controlled and capable process at minimum cost.

ADVANTAGES OF A CONTROLLED AND CAPABLE PROCESS

There are many reasons to strive for a controlled and capable process. The following are just a few.

1. More uniformity (less variation) between parts produced decreases the chances of producing nonconforming parts.
2. Fewer samples are necessary to judge product quality because they are more uniform.
3. Inspection costs are reduced, especially the cost of inspection at the end of the line.
4. There is a more accurate measure of process capability so that sound business decisions can be made regarding selection of specification limits, knowledge of the yield of good parts form a process, and selection of the proper process for the job.
5. Considerably less scrap, rework, repair, and other wastes of productive time and money are spent.

THE STANDARD DEVIATION

The standard deviation (s) is a measure of variability. Just as weight may be measured in pounds and length in inches, the variation of a process is measured in terms of standard deviation units. The standard deviation takes on the same units (pounds, inches, and so on) as the measurements. The letter s is often used for the sample standard deviation, and the Greek letter σ (sigma) is used to indicate the population standard deviation. (Note that, on many calculators, the sample standard deviation is labeled s or σ_{n-1} and the population standard deviation is the button labeled σ or σ_n.)

Approximately 99.7 percent of the area under the normal curve lies between $\overline{X}$ plus and minus three standard deviations (see Figure 7.8). An example of a capable process is shown in Figure 7.9.

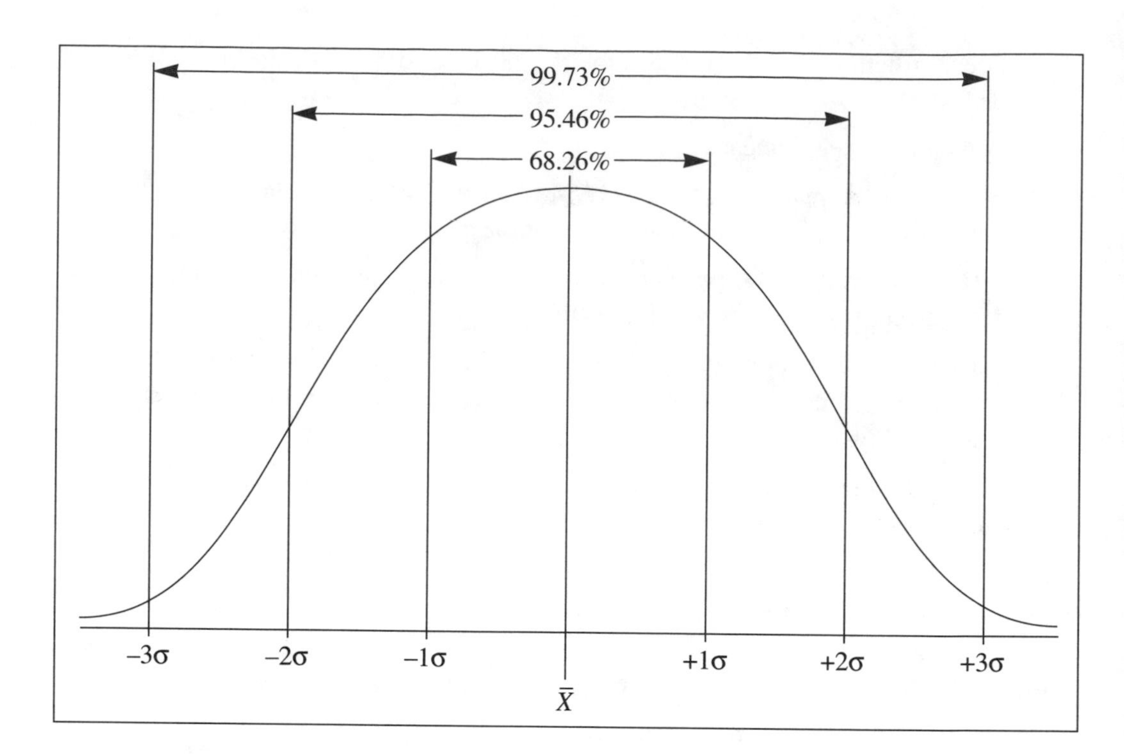

Figure 7.8. Areas under the normal curve.

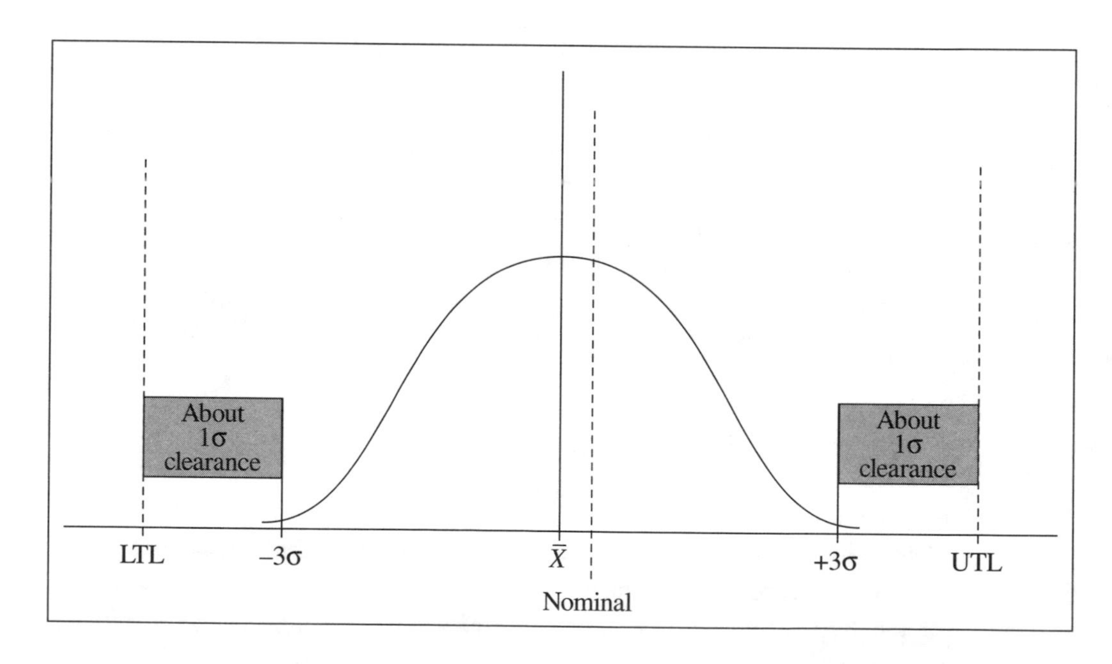

Figure 7.9. An example of a capable process.

When using control charts (such as the range chart or the sigma chart), the standard deviation of the process can be estimated by dividing the average range ($\bar{R}$) by the factor d_2 (for range charts) or dividing the average standard deviation ($\bar{s}$) by the factor c_4 (for sigma charts). Refer to Appendix B for factors. The d_2 factor depends on the sample size.

An example for a range chart is

$$\bar{R} = .003 \qquad n = 5$$

$$\hat{\sigma} = \frac{\bar{R}}{d_2} = \frac{.003}{2.33} = .0013$$

An example for a sigma chart is

$$\bar{s} = .001 \qquad n = 5$$

$$\hat{\sigma} = \frac{\bar{s}}{c_4} = \frac{.001}{.940} = .0011$$

CONTROL CHART METHOD—CAPABILITY INDEXES (C_p/C_{pk})

Steps to Evaluating Process Capability

1. Make sure the process is in statistical control.
2. Individual measurements (raw data), not averages, should be normally distributed (see Figure 7.10). This is often tested using a frequency distribution or a histogram for the individual measurements. Other tests for the normal distribution are skewness and kurtosis (not covered in this book). Note: In practice, do not expect distributions to be perfectly normal. The distribution should closely approximate a normal (bell) shape.
3. Calculate the estimated standard deviation.

$$\hat{\sigma} = \frac{\bar{R}}{d_2}$$

4. Draw lines (all to scale) on the frequency distribution that represent the grand average ($\bar{\bar{X}}$), each plus and minus three sigma limit, the nominal specification value, and the specification upper and/or lower limit (see Figure 7.11).

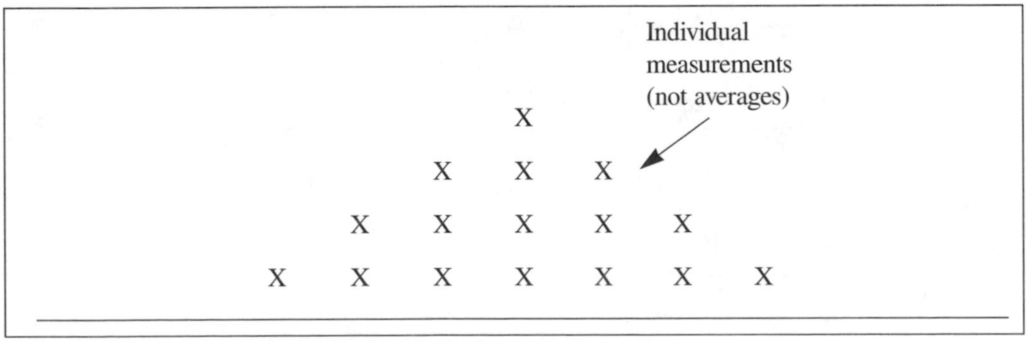

Figure 7.10. Frequency distribution applies to individuals, not averages.

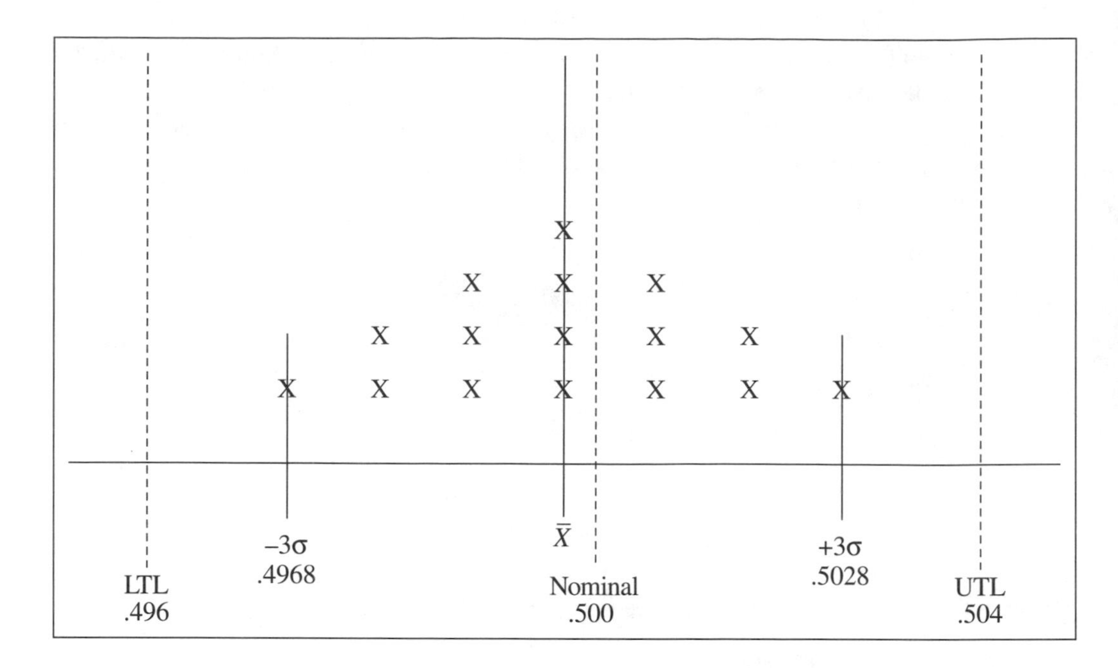

Figure 7.11. Both specification and process information is placed on the distribution.

5. Calculate the capability index (C_p) (as shown in Figure 7.11). The C_p is the tolerance width divided by the process $6\hat{\sigma}$ spread irrespective of process centering. The C_p is calculated by dividing the total specification tolerance by $6\hat{\sigma}$.

$$C_p = \frac{\text{Total tolerance}}{6\hat{\sigma}} = \frac{.008}{.006} = 1.33$$

Note: The C_p should not be less than the applicable goal. A C_p of 1.33 indicates that the process spread only consumes 75 percent of the specification tolerance (or the specification tolerance is 1.33 times wider than the process spread). The C_p indicates the capability of the process if the process average was centered between specification limits. If the process is not centered, the C_{pk} index applies.

C_{pk} Index. When the process is not centered on specification limits, a C_{pk} index applies. The C_{pk} index provides a worst-case capability index that shows the capability of the process at the time the samples were taken. This index accounts for process centering error by relating to the distance between the process mean ($\bar{\bar{X}}$) and the closest specification limit divided by one-half of the process spread ($3\hat{\sigma}$).

$$C_{pk} = \text{(the lesser of)} \ \frac{USL - \bar{\bar{X}}}{3\hat{\sigma}} \ \text{or} = \frac{\bar{\bar{X}} - LSL}{3\hat{\sigma}}$$

USL = upper specification limit
LSL = lower specification limit
$\bar{\bar{X}}$ = the grand average

For example, in Figure 7.12, the USL is closer to $\overline{\overline{X}}$, so the C_{pk} index is

$$C_{pk} = \frac{USL - \overline{\overline{X}}}{3\hat{\sigma}}$$

$$= \frac{.504 - .502}{3(.001)}$$

$$= \frac{.002}{.003}$$

$$C_{pk} = 0.67$$

Note: The C_{pk} index assumes that the process average is not centered, but it does fall on or between specification limits. If, for any reason, the process average is outside of either specification limit, the C_{pk} calculation will be a negative value.

For a process to be called capable, the C_{pk} index should be 1.33 minimum (if the goal is 1.33 minimum). This index relates to the goal of a process spread (worse case), which only consumes 75 percent of the specification tolerance.

C_p/C_{pk} Goals. The goal for process capability (using C_{pk} as an index) depends on the company's capability goals and/or customer requirements. In many companies today, a goal of 1.33 minimum exists. This means that, in the worst case, the ratio of specification tolerance (numerator) to process spread (denominator) is 4/3. Some companies have higher goals such as 1.67 or 2.0 C_{pk} minimum levels. Note that a 2.0 C_p/C_{pk} goal means 2/1 ratio.

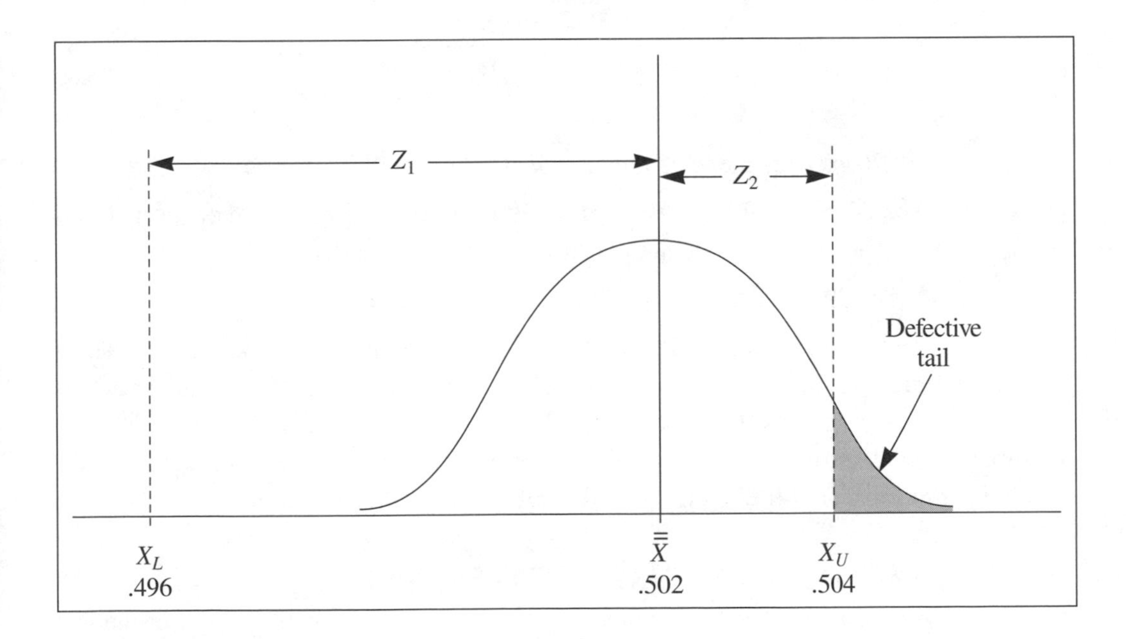

Figure 7.12. An off-centered yet capable process.

PREDICTING THE PERCENT YIELD OF A PROCESS

The yield (or percent of good parts), regardless of whether the process is centered or not, can be predicted using the table of areas under the normal curve (Appendix A). This table involves the use of a simple equation to calculate a Z score, then the Z score is used to find the area. Remember that the areas found in the normal curve table (Appendix A) are decimal fractions. You must convert these areas to a percentage. Once the Z score is found, it is used to find the area under the curve.

For example, a Z score is calculated to be 2.6. The area related to a Z score of 2.6 is .4953 (see table in Appendix A). The decimal .4953 represents 49.53 percent, which means 49.53 percent of the product is expected to fall between specification limit and the grand average.

Referring to Figure 7.12, there are two Z scores to compute. The Z score is the difference between one of the specification limits (X) and the grand average ($\bar{\bar{X}}$) divided by one standard deviation.

$$Z \text{ upper} = \frac{\text{USL} - \bar{\bar{X}}}{\hat{\sigma}}$$

$$= \frac{.504 - .502}{.001}$$

$$= 2$$

A Z value of 2 in the table represents .4772 or 47.72 percent.

$$Z \text{ lower} = \frac{\bar{\bar{X}} - \text{LSL}}{\hat{\sigma}}$$

$$= \frac{.502 - .496}{.001}$$

$$= 6$$

A Z value of 6 exceeds the table therefore use 5.0 or 49.9997 percent.

These two areas added (47.72 percent and 49.9997 percent) represent the expected percentage of good parts (within specification limits) from the process.

$$49.9997 \text{ percent} + 47.72 \text{ percent} = 97.72 \text{ percent yield}$$

This percent yield (97.72 percent) is the probability of good parts that can be expected if the process and specification do not change. The probability of nonconforming parts is equal to 100 percent minus 97.72 percent or 2.28 percent in this case. Note that, because the distribution of the process is off centered where the tail is over the USL, the 2.28 percent nonconforming are all expected to be over the USL.

THE REASON PROCESSES MUST HAVE CLEARANCE

All normal distributions have a six-sigma spread that represents 99.73 percent of the distribution. From the center of the distribution (the average), there are three standard deviations to the right and three standard deviations to the left (see Figure 7.13).

If the process distributes normally and the standard deviation is computed, one can accurately predict what the process is able to produce from a given setup ($\bar{\bar{X}}$).

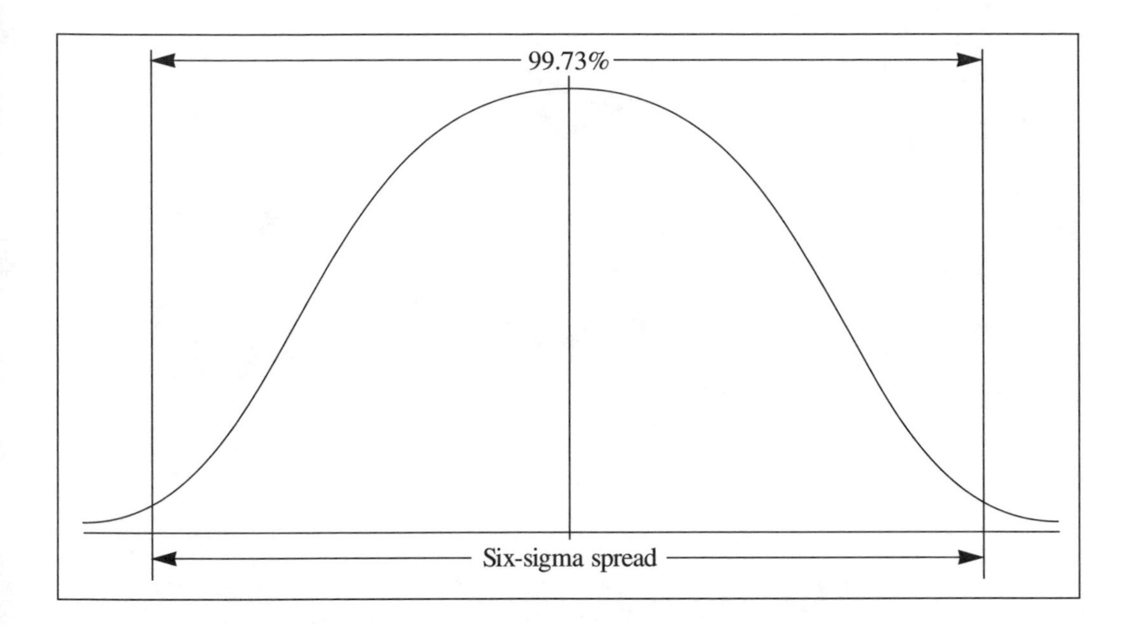

Figure 7.13. The six-sigma spread (plus and minus three sigma from the mean).

For example, where

Sigma = .001″ $\bar{\bar{X}}$ = .500″ Part specification: .503–.497

The process is in control and the distribution of the raw data is normal (see Figure 7.14). If the process is set up (or targeted) at .500, then 99.73 percent of the product will be produced within specification if the standard deviation is .001.″ The natural tolerance of a process that is in control is six sigma ($6\hat{\sigma}$), which, in this case, is 6 × .001 = .006.

Natural Tolerance

The natural tolerance refers to the inherent variation (six-sigma spread) of the process. It is what the process can do unless action is taken for process improvement. The natural tolerance has nothing to do with the specification tolerance (other than the fact that you hope it is less than the specification allows). If the natural tolerance (or the ability of the process) was equal to the specification tolerance, there would be no room for setup error or external variation such as variation in materials. This relationship is shown in Figure 7.15.

If the setup ($\bar{\bar{X}}$) moves even slightly, the curve, or process variation, moves with it. Therefore, any variation in the setup in this case would cause defectives to be produced.

To solve this problem, extra standard deviations of distance are included at each end of the curve to allow room for the setup to vary without the probability of making defective parts (see Figure 7.16). Remember, the specification tolerance is equal to the upper specification limit minus the lower specification limit. The natural tolerance is equal to six sigma (6σ) of the process. When the six sigma spread is only 75 percent of the specification tolerance and the process is centered, that extra standard deviation of room is at each tail of the curve. Six sigma can be estimated by dividing the average range ($\bar{R}$) by the d_2 factor and then multiplying by six.

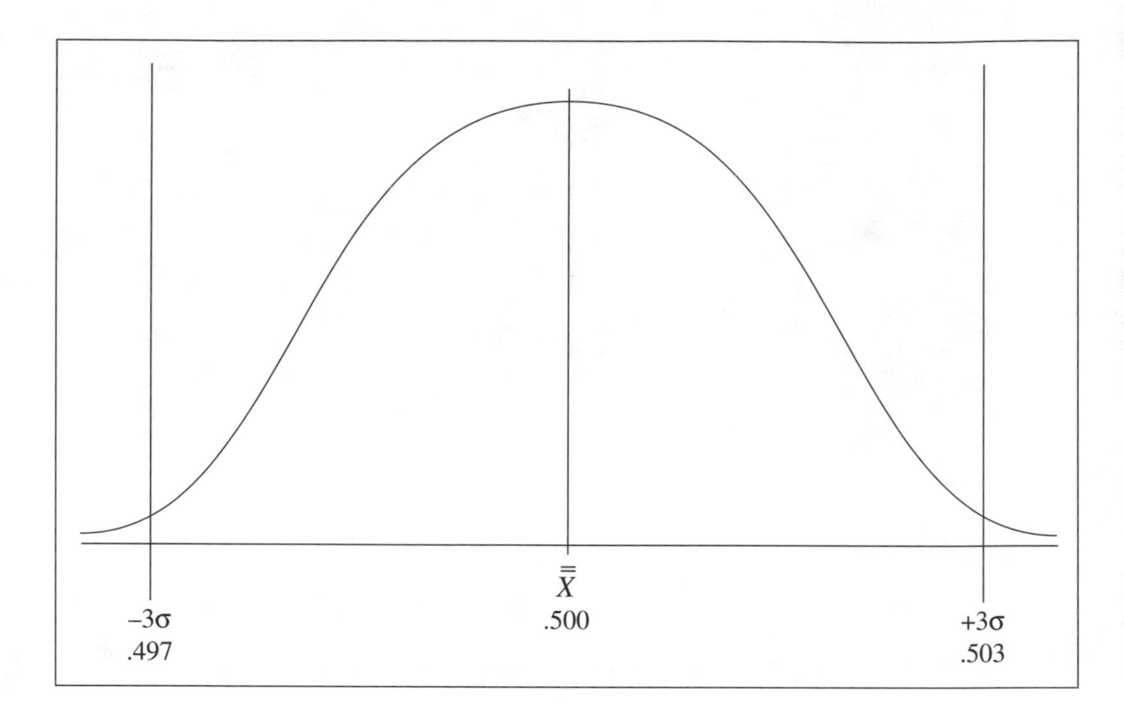

Figure 7.14. A process in control, normally distributed, and centered on nominal.

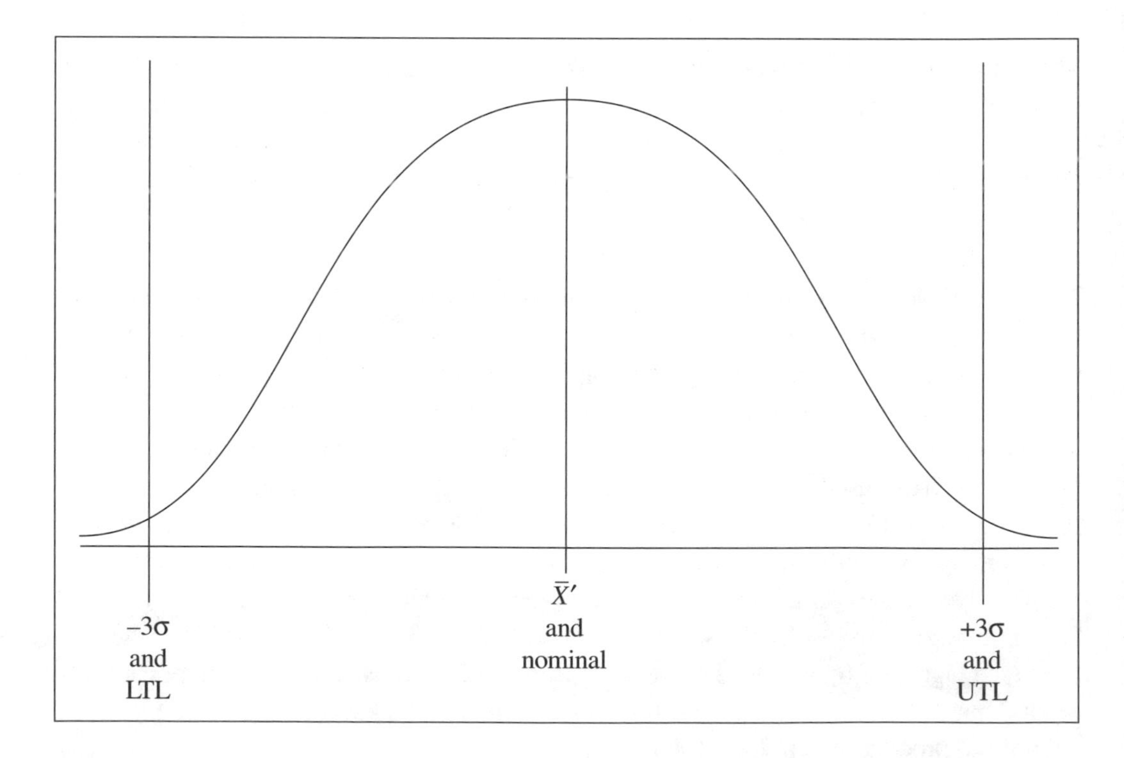

Figure 7.15. The process six-sigma spread is equal to the tolerance band.

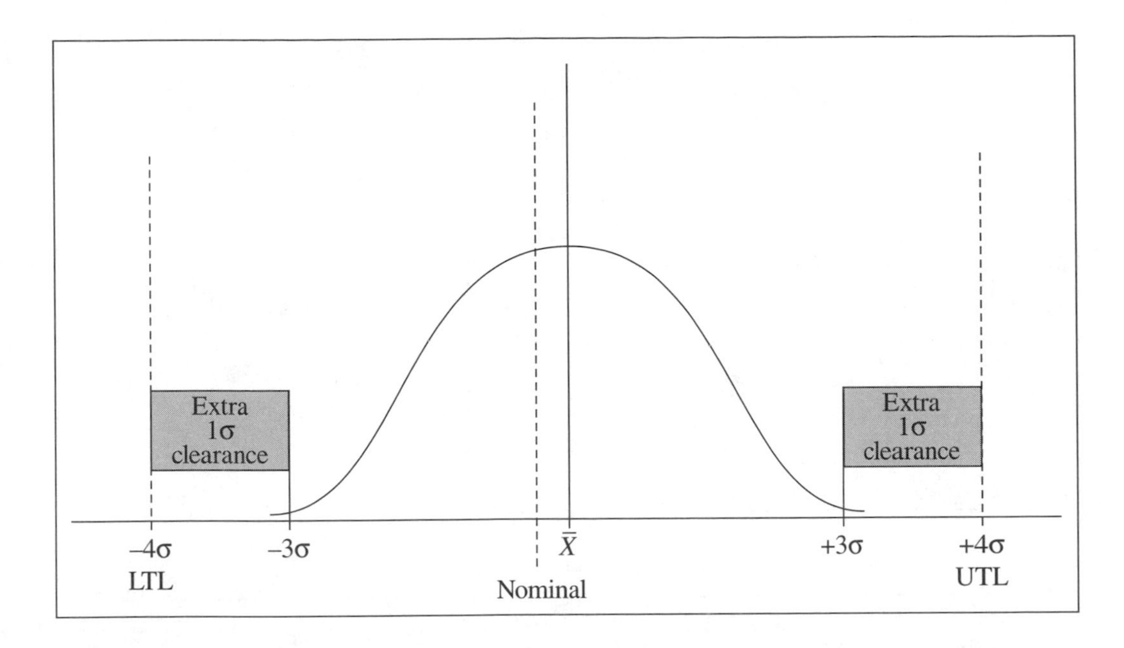

Figure 7.16. Extra clearance at the tails of the distribution.

SINGLE SPECIFICATION LIMIT

There are standards that only reflect a single limit of tolerance, such as geometric tolerances (runout, concentricity, parallelism, flatness, straightness, and so forth) and maximum/minimum limits (.500 max., .375 min., finish AA32 max.). In these cases, process capability concerns itself only with the single limit. Therefore, a Z score is computed between the only specification limit and the average of the process in order to find the percent of product that is expected to be between the average and the single limit. The other area will represent 50 percent good product in most cases (see Figure 7.17).

For example, say specification: Runout within .010 total indicator reading (TIR)

$$\text{Grand average } (\bar{\bar{X}}) = .008 \text{ TIR}$$

$$\hat{\sigma} = .0006$$

The upper Z score is 3.33 and the area under the normal curve for a Z of 3.33 is .4996 or 49.96 percent. Therefore, 49.96 percent of the product will be between .008 and .010 TIR and the other 50 percent of the product will be less than .008. Thus, the yield is 50 percent + 49.96 percent = 99.96 percent good parts. Note: The capability index (C_p) does not apply to single specification limits. Use the C_{pk} index or the areas under the normal curve. Also note that, in all cases of single specification limits, the C_{pk} is computed using the only specified limit.

C_{pk} Index

The traditional C_{pk} computation was designed for bilateral specifications where there are two limits of specification that can be violated. In bilateral cases, the C_{pk} is the smallest answer between the following two equations.

$$C_{pk} = \frac{USL - \bar{X}}{3\hat{\sigma}} \quad \text{or} = \frac{\bar{X} - LSL}{3\hat{\sigma}}$$

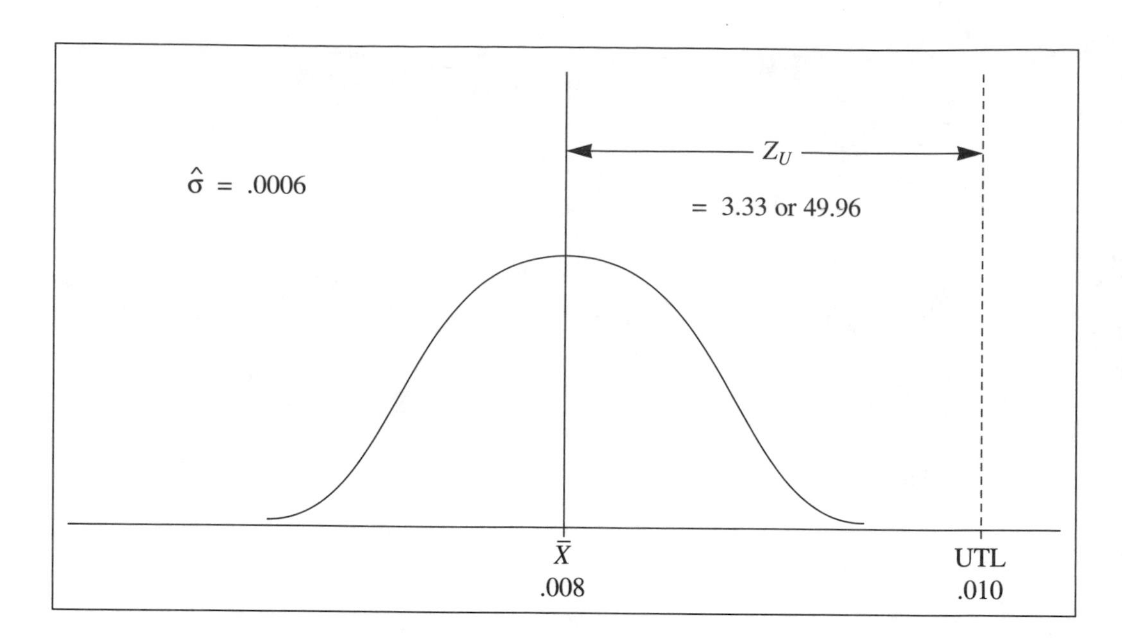

Figure 7.17. A process compared to a single specification limit.

In cases of single limit specifications, such as the one in this example, the C_{pk} is computed using only the single specified limit. The equation for maximum specification limits (and the answer to this problem) is as follows:

$$C_{pk} = \frac{USL^* - \bar{\bar{X}}}{3\hat{\sigma}} = \frac{.010 - .008}{3(.0006)} = 1.11$$

*Only the maximum specified limit was used.

If the single specification limit was a minimum limit, the C_{pk} is computed as follows:

$$C_{pk} = \frac{\bar{\bar{X}} - LSL}{3\sigma}$$

In this case, only the minimum specification limit is used.

POTENTIAL CAPABILITY STUDIES

The potential capability of a process can be studied by taking a relatively small sample of continuously produced parts, measuring them, and applying basic statistics to get an idea of the potential capability of the process. The process potential study is a short-term study that does not give you the long-term accuracy of the control chart method, but does give you an idea of what the process can do at one point in time. Potential studies have been useful in identifying the processes that are candidates for long-term process control.

Steps in Performing a Potential Capability Study

To perform a potential capability study, follow these steps.

1. Identify the process to be studied.

2. Inform the operator of the objective; it is the process that is being studied, not the person.
3. Have the process set up as close to the specification nominal as possible. This is not a must, but it does simplify the study.
4. Allow the process to run a consecutive number of parts (a minimum of 30 parts or more are recommended) without adjustments, if possible. If adjustments are necessary, record them for later use.
5. Each part should be numbered in the sequence in which it came from the process (in case further analysis is needed).
6. Accurately inspect the parts and record sizes on a data sheet next to the serial numbers.
7. Plot the data on a frequency distribution and check it for normality. The process should approximate a normal distribution for the study to be accurate.
8. Draw a nominal line and specification limits on the frequency distribution.
9. Calculate the average ($\bar{X}$) and the standard deviation (s).
10. Draw the average ($\bar{X}$) on the frequency distribution.
11. Calculate the capability index (C_p) if the process is centered or the C_{pk} index if it is not centered.
12. For a potential study the C_p and C_{pk} should be 1.33 or greater.

Example of Potential Capability Study

The example data in Table 7.2 use only 10 parts in order to simplify the example. It is recommended for actual practice that a minimum of 30 parts is used. Now the information necessary to calculate the potential capability of the process has been found. At this point, one can easily see that the process is in trouble because the C_p index is

$$C_p = \frac{\text{Total tolerance}}{6s} = \frac{.02}{6(.0059)} = 0.56$$

Table 7.2. Raw data.

Tolerance limits	Measurements			
Diameter .192–.202	(1)	.195	(9)	.196
	(2)	.196	(10)	.199
	(3)	.199	(11)	.197
	(4)	.198	(12)	.197
	(5)	.194	(13)	.200
	(6)	.196	(14)	.198
	(7)	.195	(15)	.197
	(8)	.197	(16)	.198

To compute the actual percent yield of this process involves finding the Z score for each side of the distribution. It helps to draw a normal curve and enter the process and specification data on it when calculating Z scores (see Figure 7.18). Remember, the Z score is the distance between one specification limit and the average of the process divided by the standard deviation.

$$Z = \frac{X - \overline{X}}{s}$$

There is a Z score to be calculated between the upper specification limit and the average and another between the lower specification limit and the average.

$$Z \text{ upper} = \frac{X - \overline{X}}{s}$$

$$= \frac{.51 - .5016}{.0059}$$

$$= \frac{.0084}{.0059}$$

$$Z \text{ upper} = 1.42$$

$$Z \text{ lower} = \frac{\overline{X} - X}{s}$$

$$= \frac{.5016 - .49}{.0059}$$

$$= \frac{.0116}{.0059}$$

$$Z \text{ lower} = 1.97$$

The Z scores can now be used in the table of areas under the normal curve (Appendix A) in order to find the percent expected to be between these limits.

Z upper = 1.42 ⟶ table = .42220 or 42.220 percent

Z lower = 1.97 ⟶ table = .47560 or 47.560 percent

Process yield = 89.78 percent good parts

For a process to be considered capable, the yield should be 99.994 percent (or higher). In this case, the machine is expected to produce about 90 percent good parts, and therefore 10 percent of the parts will be out of specifications.

Potential Capability Study (Practice Problem 1)

The sample data and information in Table 7.3 have been collected from a process.

Process: Milling machine

Characteristic: Thickness

Specifications: .750 ± .010

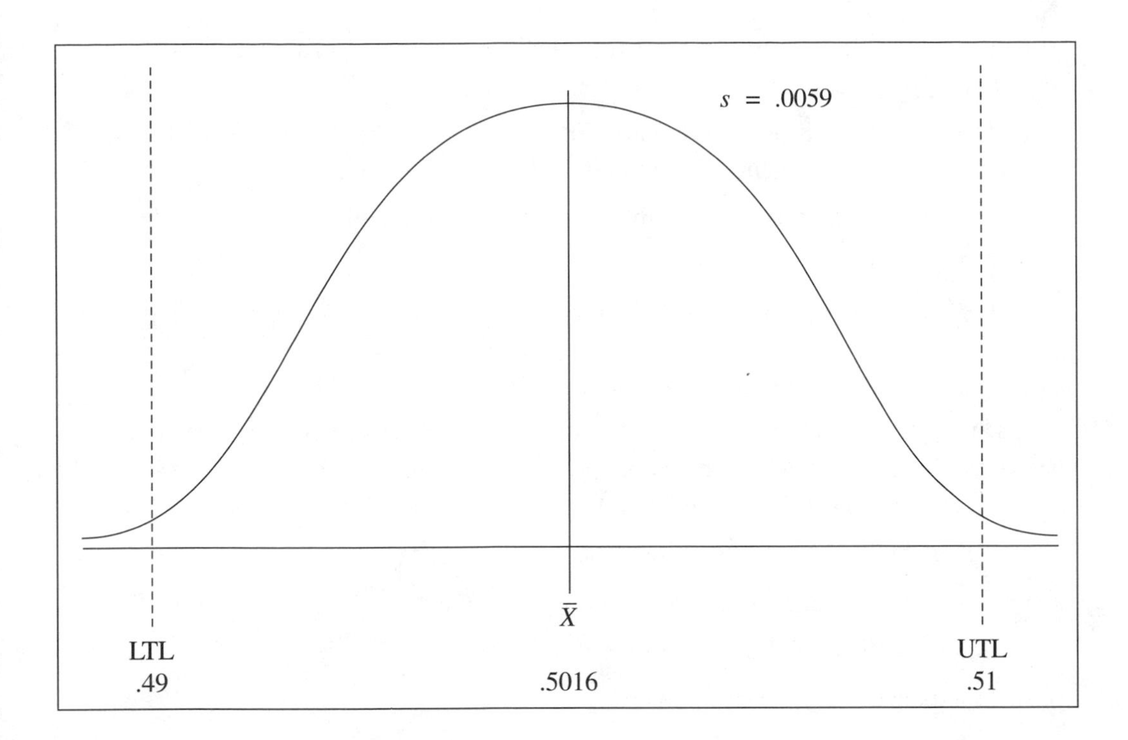

Figure 7.18. A normal curve with process and specifications compared.

Table 7.3. Machine data.

Part No.	Actual	Part No.	Actual	Part No.	Actual
1	.750	17	.756	34	.750
2	.750	18	.754	35	.750
3	.751	19	.755	36	.752
4	.751	20	.753	37	.751
5	.750	21	.751	38	.752
6	.752	22	.752	39	.753
7	.752	23	.750	40	.751
8	.753	24	.748	41	.754
9	.754	25	.749	42	.755
10	.753	26	.749	43	.754
11	.755	27	.749	44	.753
12	.758	28	.746	45	.752
13	.757	29	.747	46	.753
14	.756	30	.745	47	.752
15	.754	31	.748	48	.752
16	.755	32	.749	49	.751
		33	.751	50	.753

Answer the following questions using the worksheet provided in Figure 7.19. Refer to Appendix H for the correct answers.

a. Construct a frequency distribution and draw lines on the distribution that represent the specifications (nominal, USL, and LSL), and the process average ($\bar{X}$).

b. Calculate the average ($\bar{X}$) and the sample standard deviation (s) for the data.

c. Calculate the C_p and C_{pk} indexes.

d. Calculate the upper and lower Z scores, and then the percent yield for the process.

Potential Capability Study (Practice Problem 2)

Some sample data (in millimeters) have been taken from a process (shown in Table 7.4). A frequency distribution and histogram have been prepared for the data (as shown in Figures 7.20 and 7.21, respectively). The specification limits for the product are LSL = 3.23 and USL = 3.38.

a. Calculate the average ($\bar{X}$) of the data.

b. Calculate the sample standard deviation (s) of the data.

c. Calculate the C_p index

d. Calculate the C_{pk} index

e. Calculate the percent yield of the process.

Refer to Appendix H for the correct answers.

SHORT-RUN PROCESS CAPABILITY STUDIES—TARGET CHARTS

Process capability studies using short-run target control charts require some basic decisions before the studies are completed. Keep in mind that short-run target charts allow plotting of several different part numbers, nominal dimensions, and tolerances on one chart (for process control purposes). Process capability studies depend largely on the actual tolerance of the part compared to the variation of the process.

Short-Run Capability Studies—Methods

There are four basic methods (or approaches) to a short-run capability study. Which one of the methods used, of course, depends upon the amount of information required. The four methods are explained as follows:

• *Worst-case study* is a study that reviews all of the part numbers on the chart and compares the distribution of the process to the tightest tolerance of all the parts. Using this method, one can say that if the process is capable of the tightest toleranced part, then the capability of all other parts is better. Most manufacturers will prefer this approach because they are concerned about making all of the different part numbers to print the first time.

• *Specific-case study* is a study that isolates any particular part number and computes the capability of that part number compared to the variability of the process. This method is required to give specific customers the information they need about their specific product made on your process.

• *Every-case study* is a study that computes the appropriate capability indexes on every part number and tolerance that has been charted together on the short-run target chart.

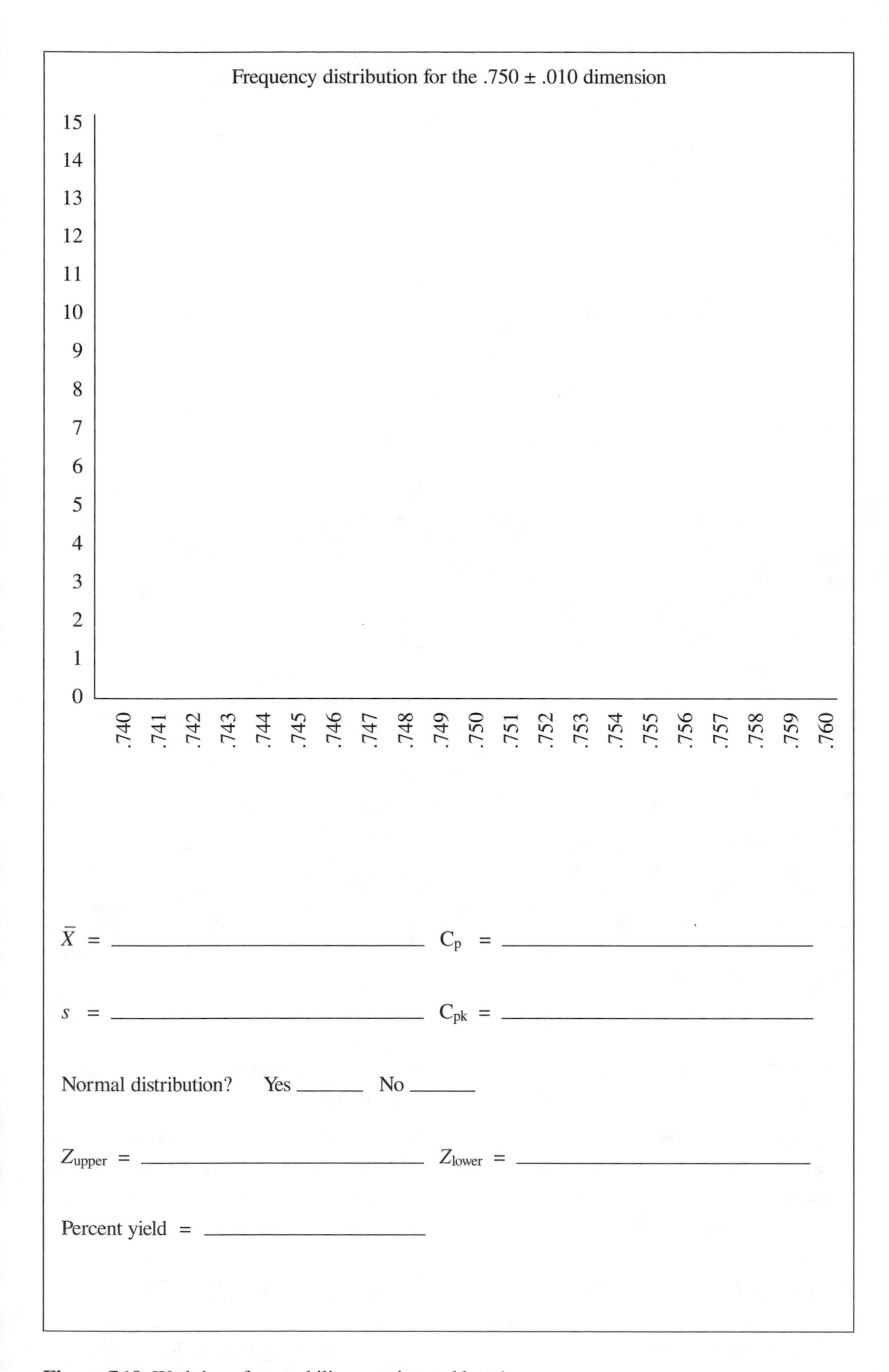

Figure 7.19. Worksheet for capability practice problem 1.

Table 7.4. Raw data (in millimeters).

3.36	3.31	3.36	3.33	3.37
3.32	3.32	3.33	3.34	3.36
3.35	3.38	3.34	3.34	3.35
3.38	3.35	3.37	3.35	3.38
3.40	3.34	3.32	3.36	3.34
3.34	3.33	3.36	3.37	3.33
3.33	3.33	3.34	3.35	3.34
3.35	3.35	3.38	3.38	3.33
3.37	3.37	3.35	3.33	3.34
3.34	3.30	3.35	3.34	3.35

				X						
				X	X					
				X	X					
			X	X	X					
			X	X	X					
			X	X	X					
			X	X	X	X	X	X		
			X	X	X	X	X	X		
		X	X	X	X	X	X	X		
		X	X	X	X	X	X	X		
X	X	X	X	X	X	X	X	X		X
3.30	3.31	3.32	3.33	3.34	3.35	3.36	3.37	3.38	3.39	3.40

Table 7.20. Frequency distribution for the data in millimeters.

This method, in most cases, is not as popular as the worst-case and specific-case studies because more effort is required to compute capability indexes.

- *Query-case study* is a study that is designed to compare a proposed tolerance change to the process or compute the actual tolerance that would be required in order for the process to be capable. This method is popular for evaluating proposed changes and the amount of change required for improvement.

USE OF A CODED FREQUENCY DISTRIBUTION (OR HISTOGRAM)

Process capability studies on short-run target charts must be performed in coded form or absolute form. In other words, the coded data on the chart do not have to be uncoded for the capability study if the values of the nominal and specification limits are also coded.

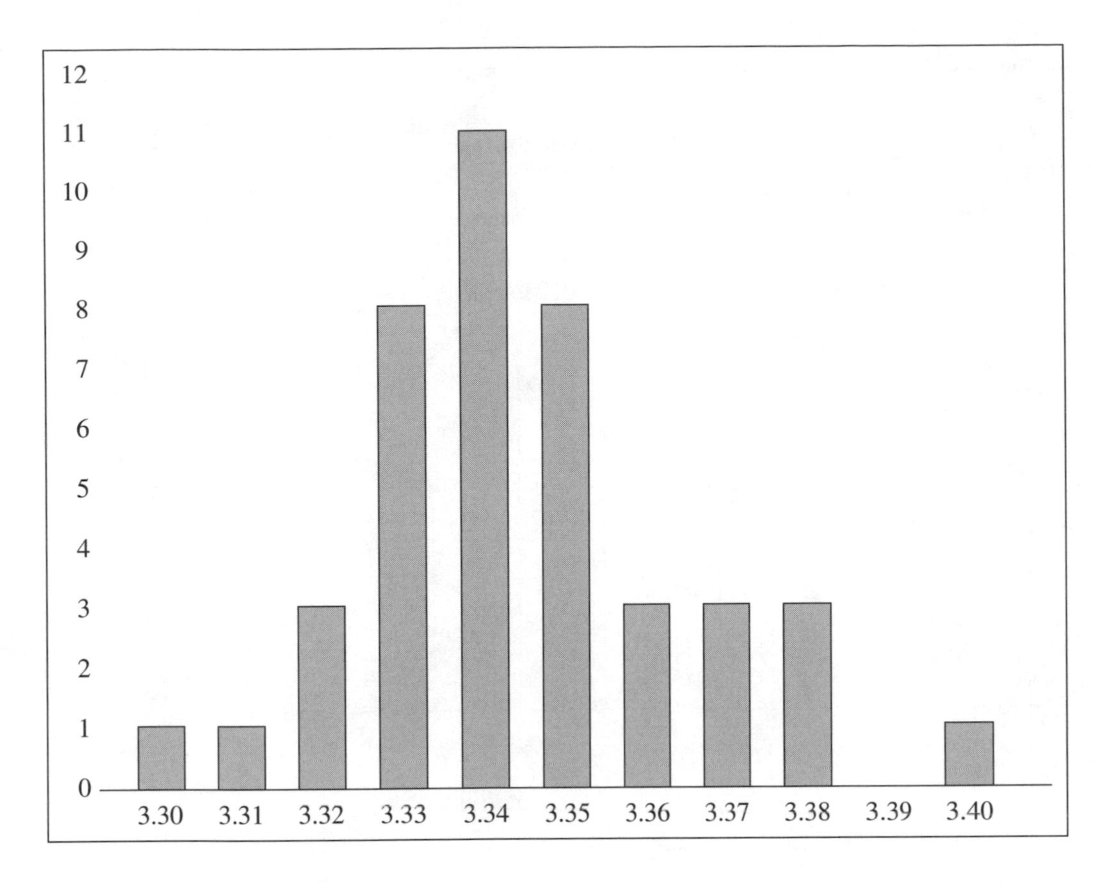

Figure 7.21. Histogram for the data (in millimeters).

Conversely, the data from the chart can be uncoded and compared to the absolute nominal and specification limits for the study. To conduct a process capability study in coded form, the distribution of coded data can be used. Figure 7.22 shows a histogram of the raw data from a short-run chart in coded format.

Using this histogram, with the coded average (–.002″) and the standard deviation (.001″), all one has to do is to code the specification target and limits using the same base from which the data were coded.

For example, if the coding was based on the nominal = zero method as shown in this example, then the coded average is compared to zero (the nominal) for central tendency purposes, and the coded specification limits are used with the coded average and the standard deviation to compute the C_p and C_{pk} ratios.

Using the following information, practice computing the C_p and C_{pk} for coded data and coded part specifications, then compare your results to the following answers.

Coded statistics (from histogram):

$\bar{\bar{X}} = -.002 \quad \sigma_{n-1} = .001$

Coded specification limits:

Nominal dim. = 0 USL = +.010 LSL = –.010

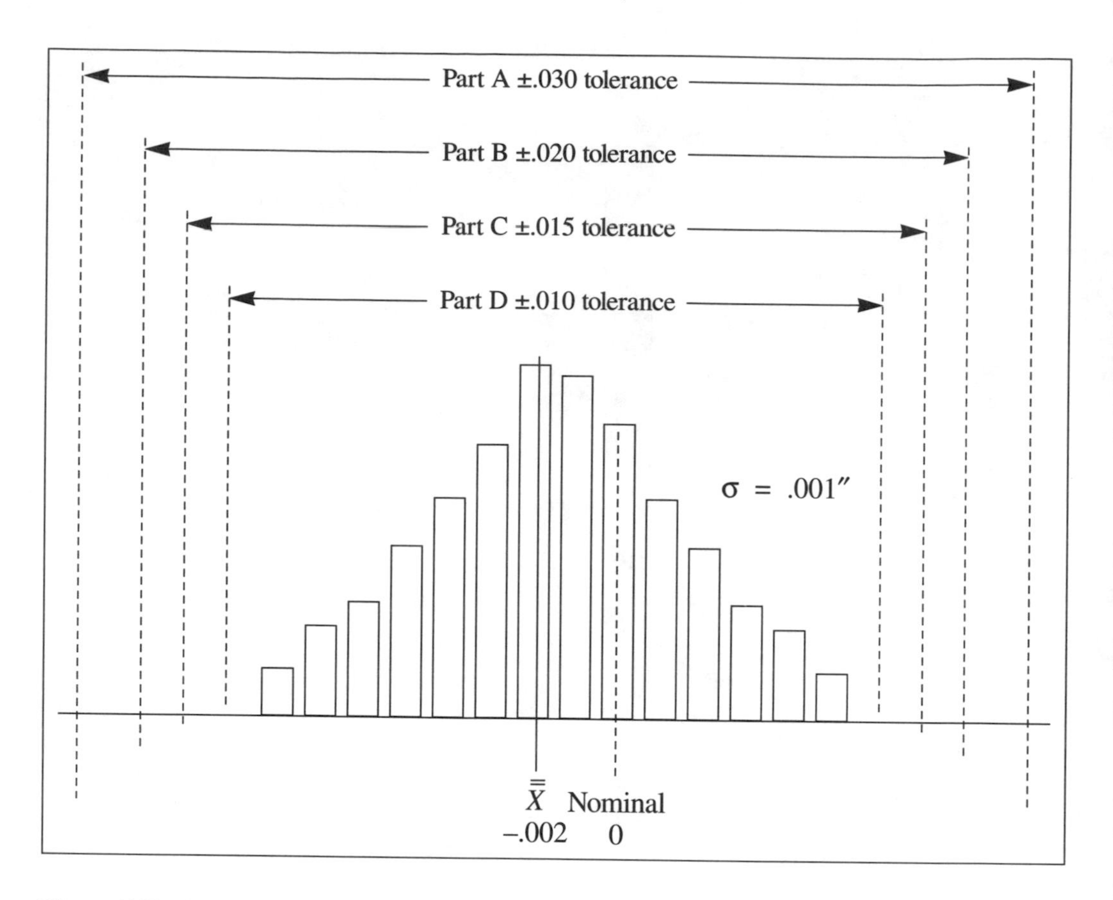

Figure 7.22. An example of a coded histogram.

Answers:

$C_p = 3.33 \quad C_{pk} = 2.67$

Table 7.5 shows an example of an every case capability study where the coded distribution of a process that is in statistical control is compared to each of five different part numbers that were charted. Refer to the coded distribution in Figure 7.22 for this problem.

The capability study method used in Table 7.5 was the every-case study. As one can see in the table, each different part number, of course, has a different C_p and C_{pk} because they all have different tolerances. Notice, however, that all of the C_p and C_{pk} values are acceptable (if the C_p/C_{pk} goal is 1.33 minimum). A worst-case study could have been used looking only at part number 123-1 because it has the tightest tolerance. The value of 2.67 C_{pk} for the tightest tolerance shows clearly that all other part numbers are better.

PROCESS CAPABILITY—ATTRIBUTE CHARTS

When an attribute chart is in statistical control, the capability is computed using the process average. For example, $\bar{P}$ is the process average for a *p* chart. Process capability, with attribute charts, reflects the yield of good product the process is expected to produce as long as the process remains in statistical control. For example, a *p* chart is in control, and the process average ($\bar{P}$) is 0.08. This means that the process is expected to produce a

Table 7.5. An example of an every-case coded capability study.

Process $\bar{\bar{X}}$	Process σ	Part number	Coded nominal	Coded tolerance	C_p	C_{pk}
–.002	.001	123-1	0	±.010	3.33	2.67
		789-3	0	±.015	5.0	4.33
		1011-4	0	±.020	6.67	6.0
		1213-5	0	±.030	10.0	9.33

proportion nonconforming (defective) of 0.08 (or 8 percent). The capability, then, is a yield of 92 percent good product.

One way the capability of an attribute chart can be improved is to apply Pareto analysis. Pareto analysis separates the vital few defects (or problems) from the trivial many. Pareto analysis identifies the major contributor(s) to the average level of poor performance of the process ($\bar{P}$) so that action can be taken to significantly reduce that level. When using attribute charts, records should be kept that identify all specific nonconformances (defects) that caused the rejection of the product so that these nonconformances can be subjected to Pareto analysis. Refer to chapter 10 on problem solving for examples of Pareto analysis. A potential capability study can be performed for attributes by using past data (from records) and excluding points that are out of control.

REVIEW PROBLEMS

Refer to Appendix H for answers.

1. A process is in control and the following data are known.

   $\bar{R}$ = .005 n = 5 $\bar{\bar{X}}$ = .500 Specification limits = .494–.506

   Calculate the capability index (C_p) for the process.

2. Is the C_p for question 1 acceptable?

   (Yes or no)

3. A process is in control, but off center with respect to the specifications. Find the C_{pk} index using the following information.

   $\bar{\bar{X}}$ = 18 PSI $\bar{R}$ = 10 PSI n = 5 Specification limits = 14–26 PSI

4. Is the C_{pk} index in question 3 acceptable? (Yes or no)

5. The C_{pk} index for question 3 can be improved if

   a. The process is centered

   b. The variation is reduced

   c. The specification limits are widened

   d. All of the above

6. Is a capability index (C_p) meaningful when a process is not centered on specification limits? (Yes or no)

7. What is the area under the normal curve for a Z score of 2.34.
8. Find the C_p index for the following data.

   $\bar{\bar{X}} = 10$ lb. $\sigma = 1$ lb. Specification limits = 8–12 pounds
9. Which of the following actions can be taken to improve the C_p in question 8?
   a. Adjust (center) the process
   b. Set up the process toward the high limit
   c. Change the upper specification limit
   d. Reduce the variation in the process
10. If a process is in control using a p chart, and the average fraction defective is .15, what is the capability of the process?
11. A short-run chart is being maintained for the following parts and their associated tolerances. The process was in statistical control during the production of all part numbers. The grand average of the process is +.001 and the standard deviation is .0005." What is the worst-case C_{pk} index?

Part	Nominal	Tolerance
A	.500	± .010
B	.750	± .005
C	.375	± .020

12. Is the process in question 11 capable of producing all of the part numbers shown?
13. In question 11, if a new part number (D) with a tolerance of ±.002 is produced on this process, would the process be capable of producing this part to a 1.33 minimum C_{pk} requirement?
14. What is the C_p index for part A in question 11?
15. Using the knowledge of the process in question 11, what action could be taken on the process that will increase all of the C_{pk} values?

CHAPTER 8

Gage Repeatability and Reproducibility (R&R) Studies

INTRODUCTION

SPC is a powerful tool for controlling processes and making the process capable of high first-time yields of good product. It is understood, however, that SPC analysis is only as good as the accuracy and reliability of the measurements. This chapter covers gage R&R analyses methods that are designed to measure the capability of the measurement system (the observer and the measuring instrument). The variability of the measurement system is the ability of the observer(s) and the instrument to obtain accurate and repeatable measurements. Accurate and precise measurements help to avoid false product acceptance decisions and false signals on control charts (such as the flat-line pattern shown in Figure 8.1).

Before getting into gage R&R studies, let's first review the definitions of common terms.

Accuracy is the difference between the average of measured values on a part and the true value of the part.

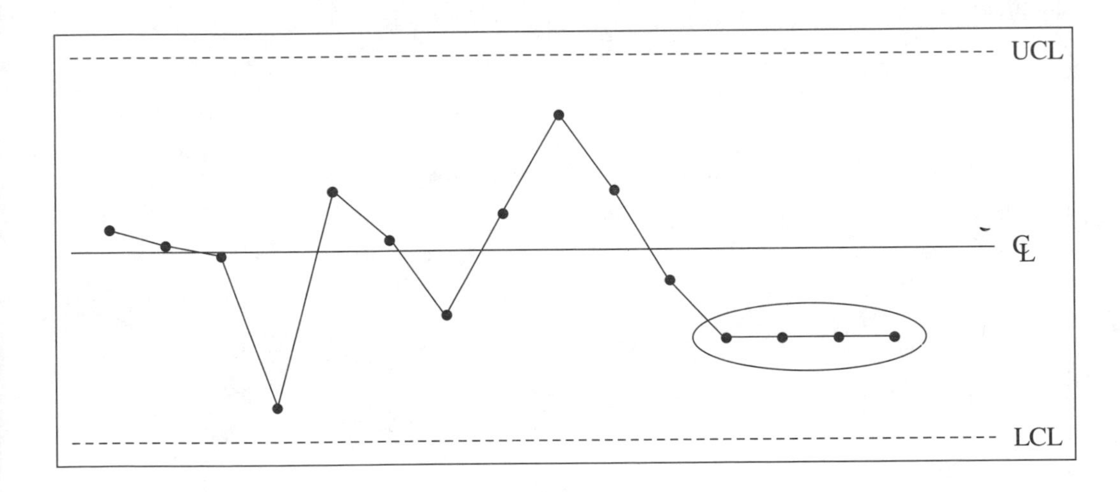

Figure 8.1. A flat-line pattern occurs due to gage error.

Discrimination is the finest line graduation of the measuring instrument (for example, the discrimination of a vernier micrometer is .0001," a steel rule is 1/64," and so on).

Gage R&R results are shown in terms of percent of part tolerance consumed by the variability of the measurement system.

Measurement system refers to the observer and the equipment.

Precision means getting consistent results repeatedly.

Repeatability is the ability of one observer to obtain consistent results measuring the same part (or set of parts) using measuring equipment.

Reproducibility is the overall ability of two or more observers to obtain consistent results repeatedly measuring the same part (or set of parts) using the same (or similar) measuring equipment.

Ten percent (10%) rule states the discrimination of the measuring instrument should not exceed 10 percent of the total part tolerance. For example, if a part has a total tolerance of .010," the discrimination of the selected measuring instrument should not exceed 10 percent of .010" or .001."

Accuracy is the ability to obtain a true value (hit the target). *Precision* is the ability of the instrument to repeat its own measurements. Many people confuse the two or consider them the same thing. Instruments can be inaccurate and precise (meaning they repeat the measurement but the measurement is in error). For example, a micrometer may be dropped on the floor and the frame becomes bent. This micrometer will repeatedly be in error. Precision problems with measuring instruments can only be improved by identifying and correcting the source of instrument variation.

Consider the rifle range results of three shooters firing five shots each as shown in Figure 8.2. Accuracy is the ability of the shooter to hit the target, where precision is the ability of the shooter to repeat the results (or a tight group of target hits). Shooter A is accurate. All of the bullets hit the target. Shooter B is precise. All of the bullets are tightly grouped together (even though the shooter missed the target completely). Shooter C is both accurate and precise.

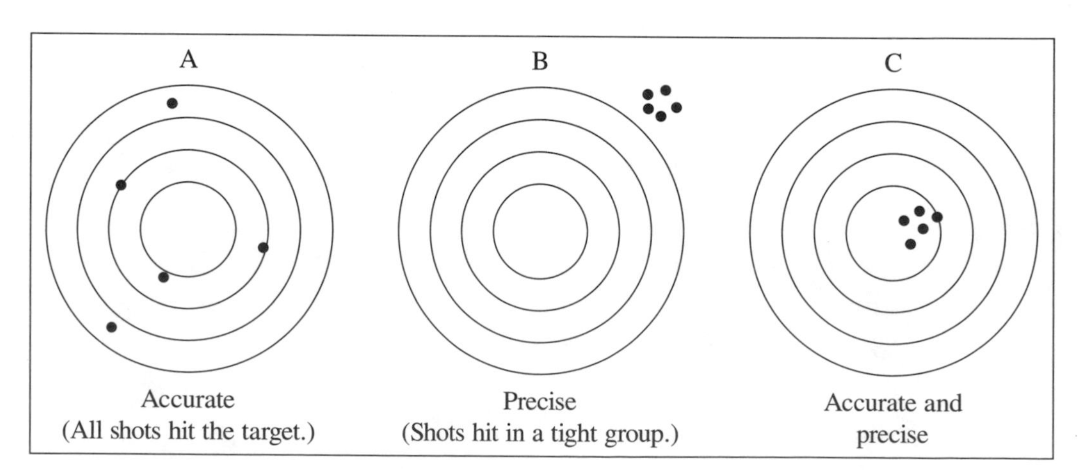

Figure 8.2. Accuracy and precision on the rifle range.

When measuring products for product acceptance (inspection) or process control purposes, it is important that the measurements are accurate and precise in order to avoid measurement error.

MEASUREMENT ERRORS

There are two types of risks involved with measurement errors. In product acceptance, these risks are called *alpha risk* and *beta risk*. In process control, they are referred to as the *type I* and *type II errors* respectively.

The alpha risk, in product acceptance, is the risk of rejection of good product due to measurement error and holding good product from further production and shipment. This is also called producer's risk.

The type I error, in process control, is the risk of taking action on a process that appears to be out of control or incapable but actually is not out of control or incapable.

The beta risk, in product acceptance, is the risk of acceptance of defective product due to measurement error and shipping the defective product to the customer. This is sometimes called consumer's risk.

The type II error, in process control, is the error of not taking action on an out-of-control or incapable process because the process appears to be in control and/or capable due to measurement error.

There are numerous cases in industry where measurement error has caused significant production, cost, inventory, and customer problems, just to name a few. The bottom line is *if you can't measure it, you don't understand it.*

There are a wide variety of measurement errors to avoid. The following examples are a few of the common types of measuring errors categorized as errors due to the observer, or errors due to the measuring equipment.

Instrument-related errors

- The equipment is out of calibration.
- The equipment has worn or broken parts.
- An indicator is sluggish or sticking.
- Inaccurate equipment setting.

Observer-related errors

- The observer has not been trained on how to use the instrument.
- The observer lacks the skill to use the instrument (for example, where feel is required).
- The observer is biased on the outcome of the measurement.
- Viewing dial indicating devices from an angle, which distorts the observer's reading of the device (parallax error).
- Discrimination (an instrument that discriminates larger than 10 percent of the part total tolerance).
- The wrong type of instrument for the measurement is chosen.
- The wrong datum references are contacted for measurement.

- Scale readings of the instrument are rounded off.
- The instrument is misheld or manipulated.

Instrument errors can be isolated using gage R&R analysis and then corrected in one of the following ways.

- Correct the calibration.
- Select a better instrument.
- Select a different instrument.
- Repair the instrument.
- Replace the instrument.
- Modify the instrument.

Observer errors can also be isolated using gage R&R analysis. In most cases, observer errors are due to the lack of training or skill with regard to the instrument being used. Once these errors are found, they can then be corrected in one or more of the following ways.

- Train the observer, then restudy the process.
- Discuss rounding/bias/and so on (and the adverse effects) with the observer.
- Make sure all observers use the same type of instrument (or the same one).

WHY PERFORM GAGE R&R STUDIES?

There are many good reasons for doing gage R&R studies, other than the fact that it is a good practice to provide some guarantee that the measurement results you're getting are true. Some benefits of gage R&R studies are

- They help to avoid significant measurement error on product acceptance and on control charts and/or process capability studies.
- They provide criteria for accepting new measuring equipment.
- Comparisons between different measuring devices can be made more effectively.
- They help to evaluate deficient gages as opposed to incapable/out-of-control processes.
- They provide validity for statistical control and capability methods used in the process.
- They improve equipment maintenance, selection, calibration, and use.
- They improve observer's training and skill (as necessary).

WHEN SHOULD GAGE R&R STUDIES BE PERFORMED?

The most effective time to perform gage R&R studies is before the gage is used to gather data or inspect products. In this preventative mode, problems that are associated with taking data, charting, capability analysis, acceptance errors, and so on are avoided because the measurement system that fails a gage R&R study is corrected before data are collected. Along with the prevention mode, however, comes the task of performing a variety of studies before data collection can begin.

Another school of thought is the reactive mode. In this mode, if control and capability problems arise that are unexplained in the process, one challenges the measurement system and performs a gage R&R study at that time. The reactive mode is not recommended because, in most cases, a considerable amount time and energy has been wasted on collecting bad data.

Gage R&R is a proven analysis method for measuring the capability of the measurement system and isolating the primary source(s) of measurement errors (observer or instrument). Using gage R&R, the intent is to (1) identify the measurement processes that require improvement, (2) identify the source of measurement errors (observer or equipment), and (3) correct the error.

PLANNING FOR THE GAGE R&R STUDY

Many gage R&R studies fail because the fundamentals of proper measurement have not been observed when the measurement process was selected. Gage R&R analysis measures the variability of the measuring process regardless of whether it has gross errors or not. The following steps will help avoid false starts and failed studies.

Step 1: Select the Proper Instrument

Make sure that the instrument selected for the measurement is the right kind of instrument to perform the measurement. For example, reference and measured contact faces are those that adequately make contact with the part reference and measured surfaces (for example, it is not proper to use a pointed micrometer on an outside diameter).

Datums are often encountered when dimensionally measuring geometric tolerances. *Datums* are points, lines, axes, or planes that must be derived from a true geometric counterpart on the part itself. If the instrument does not properly contact the part, datums will not be properly established, and there will often be appreciable measurement errors.

Step 2: Follow the 10% Rule of Discrimination

If the 10% rule of discrimination is not followed, it is very likely that the gage R&R study will fail, and the primary source will be the instrument. In fact, some gage R&R studies may actually pass with flying colors when the measurements are still in error. At times, the 10% rule is not sufficient. For example, the gage does discriminate to 10% of the part tolerance, but the process variation is so small compared to part tolerance that the gage becomes inefficient in measuring the products. In these cases, one may need to use a finer discrimination of instrument in order to see the variation of the process.

Step 3: Make Sure of the Observers' Training/Skill

At the start of selecting the measuring process, the ability/skill of the observers should at least be established via interviews or witnessing the observers using the instrument.

Step 4: Make Sure All Observers Use the Same Instrument

The same instrument, or at least the same type of instrument, should be used by all observers. (For example, all observers must use a calibrated .001″ discrimination micrometer). The assumption here is that all equipment used is calibrated.

Step 5: Calibrate the Equipment

Make sure that all equipment being used is calibrated to avoid fundamental equipment-related measuring errors.

ACCURACY MEASUREMENT

Accuracy is defined as the difference between the average ($\bar{X}$) of several measurements made on a part dimension and the true value of that dimension (often called μ [or mu]). The distance between $\bar{X}$ and μ is referred to as the *error, bias,* or *inaccuracy.* Accuracy error can be positive or negative depending on the direction of the error of the observer's measurements.

$$\text{Accuracy} = \bar{X} - \mu \text{ or } \mu - \bar{X}$$

Accuracy is affected by several types of observer-related or equipment-related measurement errors, such as the tool is out of calibration or incorrectly set, the observer has poor feel, or the wrong datum reference is used.

Some instruments are considered accurate if the inaccuracy is within the tolerance for that grade of instrument. Overall, the measuring equipment should have the ability to measure to 10 percent of the total product tolerance (or, at times, 10 percent of the process variation).

Accuracy Measurement in Gage R&R Studies

Traditional gage R&R studies, by design, are concerned with measuring and understanding the repeatability and reproducibility of the observer(s) and the equipment. The methods do not include measuring inaccuracy. In order to add accuracy measurement to the study, use known parts (either production parts with known values or standard parts made especially for the study).

Known parts are generally production parts that have been previously measured by very accurate equipment (such as parts for a gage R&R study on handheld micrometers that have been previously measured with a super micrometer). Note, however, that these parts may have within-the-piece variation, such as taper, and may have to be measured in a specific location.

Standard parts are parts that have been made to order by precision equipment (for example, precision blocks). These parts ordinarily do not have within-the-piece variation, so measurement location is not important. Along with all of the calculations performed in gage R&R studies, in this case, accuracy calculations can also be made (knowing the true value of the parts).

ACCURACY STUDIES—PRODUCTION PARTS

To determine accuracy in using production parts, complete the following steps.

1. Obtain three parts for the study.
2. Find the true value (μ) of each part by measuring each part at a specific location with a gage that discriminates to one finer decimal place than the gage to be studied.
3. Have the observer measure each of the three parts 10 times at the specified location.

4. Calculate the average ($\bar{X}$) of the observer's 10 measurements for each part. ($\bar{X}_1$, $\bar{X}_2$, $\bar{X}_3$, and so on.)
5. Calculate the accuracy error for each part.

$$\text{Acc} = \bar{X} - \mu$$

6. The accuracy, in this case, is the maximum difference. Note: If only one part was used for the study, the accuracy is found directly by the above equation.

To determine the results, calculate the accuracy percent of part tolerance (Acc%).

$$\text{Acc\%} = \frac{|\bar{X} - \mu|}{\text{Total tol.}} \bullet 100$$

Typically, if the accuracy error is ≤ 5 percent of part tolerance, it is acceptable. If the accuracy error exceeds the 5 percent (or other accuracy requirements), corrections should be made on the cause(s) of inaccuracy. Corrections for accuracy should be made using *binary search,* or correcting half of the error at a time, then retesting.

In cases where there is more than one observer, compute the accuracy (±) and the accuracy percent (Acc%) for each observer and make corrections. Also compute the accuracy (±) and accuracy percent (Acc%) for all observers using the grand average ($\bar{\bar{X}}$) of all observers.

ACCURACY STUDIES—STANDARD PARTS

To perform an accuracy study with standard parts, complete the following steps.

1. Obtain at least three standard parts of known size (for example, parts specifically ground and lapped to size within millionths).
2. Have the parts measured in a metrology lab and certify their true value (μ). Note: Parts should be made of typical gage materials for stability.
3. Have the observer measure each of the three parts 10 times. Note: Location should not be a concern with standard parts because they have no appreciable within-the-piece error.
4. Calculate the average ($\bar{X}$) of the observer's 10 measurements for each part. ($\bar{X}_1$, $\bar{X}_2$, $\bar{X}_3$, and so on.)

$$\text{Acc} = \bar{X} - \mu$$

5. Calculate the accuracy error for each part.
6. The accuracy, in this case, is the maximum difference. Note: If only one part was used for the study, the accuracy is found directly by the above equation.

To obtain results, calculate the accuracy percent of part tolerance (Acc%):

$$\text{Acc\%} = \frac{|\bar{X} - \mu|}{\text{Total tol.}} \bullet 100$$

Typically, if the accuracy error is < 5 percent of part tolerance, it is acceptable. If the accuracy error exceeds the 5 percent (or other accuracy requirements), corrections should be made on the cause(s) of inaccuracy. Corrections for accuracy should be made using binary search.

In cases where there is more than one observer, compute the accuracy (±) and the accuracy percent (Acc%) for each observer and make corrections.

GAGE R&R STUDIES—RANGE METHODS

The range method is typically used when performing gage R&R studies (especially when they are performed manually). There are other methods for gage R&R studies (not covered in this book) that use the standard deviation instead of the range. There are two popular range methods that can be used: the short form and the long form. Both of the range methods involve basic math. The drawback of both range methods is the fact that those who perform the study must use estimates of the standard deviation derived from the ranges of the data. The long form is recommended; however, for a quick look at gage R&R, the short form can provide a good estimate. Calculators or software should be used for gage R&R analysis.

The Long-Form Range Method

This practice problem will help you learn and perform the long form.

Two observers are being studied using a micrometer to measure the thickness of a part. The part thickness dimension and tolerance is .047 ± .0025″ (a total tolerance of .005″). It has been decided that the observers will measure each part three times (trials). The measurements of each observer are shown in Figure 8.3. Refer to Figure 8.3 in each case for the data and Figure 8.4 for the results of each step. The following steps can be used to perform a gage R&R (long-form) study.

Step 1. Obtain n samples (from 5 to 10 parts are recommended) from the process and number the samples 1 . . . n. These parts should represent the parts that are going to be measured for controlling the process. Note: If possible, the known value of these parts should be obtained for the purpose of studying accuracy. Typical gage R&R studies do not study accuracy, just repeatability and reproducibility. In this case, five parts have been selected for the study.

Step 2. Decide on the number of trials (how many times each part will be measured). Two or three trials are recommended. Have each observer (label observers A, B, and so on) measure all n parts two or three trials (times) each and record the measurements. In this example, five parts are being studied and three trials will be used as shown in the gage R&R worksheet in Figure 8.3. Note: Each observer measures all n parts in a row, then all n again, and so on. Observers should not be allowed to see another observer's measurements (or their own previous measurements) to avoid bias in the data. A method of collecting the data should be devised that will provide assurance that observers' measurements are unbiased.

Step 3. For each observer, calculate the average ($\overline{X}$) of each sample and record it on the worksheet. Then calculate the grand average of each observer's averages and enter the grand average on the worksheet as $\overline{X}_A$, $\overline{X}_B$, and so on.

Step 4. For each observer, find the range (R) of each sample and record it on the worksheet.

Obs.	A					B					Statistics	
Sample	Trial	Trial 2	Trial 3	Avg.	Range	Trial 1	Trial 2	Trial 3	Avg.	Range	$\bar{X}_A$ =	$\bar{X}_B$ =
1	.0473	.0472	.0472			.0470	.0469	.0472			$\bar{X}_{diff}$ =	
2	.0468	.0469	.0471			.0471	.0471	.0469			$\bar{R}_A$ =	$\bar{R}_B$ =
3	.0472	.0472	.0470			.0470	.0471	.0470			$\bar{\bar{R}}$ =	
4	.0471	.0473	.0471			.0470	.0470	.0471				
5	.0468	.0469	.0471			.0467	.0469	.0469			UCL_R =	

Equipment variation (EV)	Appraiser variation (AV)	Total R&R
$EV = \bar{\bar{R}} \bullet K_1$	$AV = \sqrt{[\bar{X}_{\text{diff}} \bullet K_2]^2 - \frac{(EV)^2}{n \bullet r}}$ n = number of parts r = number of trials	$R\&R = \sqrt{(EV)^2 + (AV)^2}$
Equipment variation percent	**Appraiser variation percent**	**R&R %**
$EV\% = \frac{EV}{\text{Total tolerance}} \times 100$	$AV\% = \frac{AV}{\text{Total tolerance}} \times 100$	$R\&R\% = \frac{R\&R}{\text{Total tolerance}} \times 100$

Figure 8.3. Gage R&R study—range method (long form) worksheet.

Obs.	A					B					Statistics
Sample	Trial	Trial 2	Trial 3	Avg.	Range	Trial 1	Trial 2	Trial 3	Avg.	Range	$\bar{X}_A = .04708$ $\bar{X}_B = .04699$
1	.0473	.0472	.0472	.04723	.0001	.0470	.0469	.0472	.04703	.0003	$\bar{X}_{diff} = .00009$
2	.0468	.0469	.0471	.04693	.0003	.0471	.0471	.0469	.04703	.0002	$\bar{R}_A = .00022$ $\bar{R}_B = .00018$
3	.0472	.0472	.0470	.04713	.0002	.0470	.0471	.0470	.04703	.0001	$\bar{\bar{R}} = .0002$
4	.0471	.0473	.0471	.04717	.0002	.0470	.0470	.0471	.04703	.0001	
5	.0468	.0469	.0471	.04693	.0003	.0467	.0469	.0469	.04683	.0002	$UCL_R = .00051$

Equipment variation (EV)

$$EV = \bar{\bar{R}} \bullet K_1 = .0002 \bullet 3.05 = .00061$$

Appraiser variation (AV)

$$AV = \sqrt{[\bar{X}_{diff} \bullet K_2]^2 - \frac{(EV)^2}{n \bullet r}}$$

$$= \sqrt{[.00009 \bullet 3.65]^2 - \frac{(.00061)^2}{5 \bullet 3}}$$

$$= .00029$$

n = number of parts
r = number of trials

Total R&R

$$R\&R = \sqrt{(EV)^2 + (AV)^2}$$

$$= \sqrt{(.00061)^2 + (.00029)^2}$$

$$= .00068$$

Equipment variation percent

$$EV\% = \frac{EV}{\text{Total tolerance}} \times 100$$

$$= \frac{.00061}{.005} \times 100$$

$$= 12.2\%$$

Appraiser variation percent

$$AV\% = \frac{AV}{\text{Total tolerance}} \times 100$$

$$= \frac{.00029}{.005} \times 100$$

$$= 5.8\%$$

R&R %

$$R\&R\% = \frac{R\&R}{\text{Total tolerance}} \times 100$$

$$= \frac{.00068}{.005} \times 100$$

$$= 13.6\%$$

Figure 8.4. Gage R&R study—range method (long form) worksheet.

Step 5. Calculate the average range ($\bar{R}$) for each observer using the ranges for all of that observer's trials. In this case, the average range for Observer A is .00022″ and the average range for Observer B is .00018.″

Step 6. Calculate the grand range ($\bar{\bar{R}}$) by averaging the $\bar{R}$s of all observers. In this case, $\bar{\bar{R}}$ = .0002.″

Step 7. Calculate the difference between the maximum observer average value and the minimum observer average value (called $\bar{X}_{diff}$). In this case, $\bar{X}_{diff}$ = .00009.″

Step 8. Calculate the upper control limit for the ranges by multiplying the overall average range ($\bar{\bar{R}}$) times the appropriate D_4 value (refer to the table of factors for control charts in Appendix B). The D_4 factor depends upon the number of trials. If any of the observers' ranges exceed the upper control limit for ranges, either repeat those readings from that observer or discard the values, go back to step 3, and recalculate. In this case, the UCL of ranges equals $\bar{\bar{R}}$ times D_4 or .0002″ times 2.574 or .00051.″ There are no observer ranges that exceed the UCL of .0005,″ so the study can be continued.

Step 9. Calculate the equipment variation (EV) using the average range and the K_1 factor. The K_1 factor also depends on the number of trials (see Table 8.1). The equation for K_1 is shown for reference only.

$$\text{EV} = \bar{R}_0 \times K_1 = .0002 \times 3.05 = .00061$$

$$K_1 = \frac{5.15}{d_2}$$

Step 10. Calculate the percent of tolerance consumed (EV%) by equipment variation.

$$\text{EV\%} = \frac{\text{EV}}{\text{Total tolerance}} \times 100 = \frac{.00061}{.005} \times 100 = 12.2\%$$

Step 11. Calculate the appraiser variation (AV). See Table 8.2 for K_2 factors.

$$AV = \sqrt{[\bar{X}_{diff} \bullet K_2]^2 - \frac{(EV)^2}{n \bullet r}}$$

$$= \sqrt{[.0009 \bullet 3.65]^2 - \frac{(.00061)^2}{5 \bullet 3}}$$

$$= .00029$$

where

n = is the number of parts

r = is the number of trials

K_2 = depends on the number of observers being studied

Table 8.1. K_1 factors.

Trials	2	3
K_1	4.56	3.05

Table 8.2. K_2 factors.

Observers	2	3
K_2	3.65	2.70

Step 12. Calculate the percent of tolerance consumed by appraiser variation (AV%).

$$AV\% = \frac{AV}{\text{Total tolerance}} \times 100 = \frac{.00029}{.005} \times 100 = 5.8\%$$

Step 13. Calculate the repeatability and reproducibility (R&R) of the measurements. Note: These calculations are based on predicting 5.15 sigma or 99 percent of the area under the normal curve.

$$R\&R = \sqrt{(EV)^2 + (AV)^2} = \sqrt{.00061^2 + .00029^2} = .00068$$

Step 14. Calculate the percent of tolerance consumed by the measurement process (R&R%).

$$R\&R\% = \frac{R\&R}{\text{Total tolerance}} \times 100$$

$$= \frac{.00068}{.005} \times 100$$

$$= 13.6\%$$

The total R&R% would be ideal if it were 10 percent or less, but each company has to establish its chosen goal of measurement capability. Typically 10 percent or less is considered very good R&R, 20 percent to 25 percent is considered marginally capable, and over 25 percent is considered unacceptable.

The Short-Form Range Method

A shorter method for performing gage R&R studies using ranges is the short-form range method. Follow this practice problem to learn how to perform a short-form range study.

A part has a dimension and tolerance of .501 ± .0025. Five parts are drawn at random, and there are two observers involved in the study. For information purposes, the parts are sequentially numbered one through five.

Step 1. Have each observer measure each part once and record the measurements on a data collection sheet (as shown in Table 8.3). Do not let the observers see each other's data to avoid bias.

Step 2. Find the range between the observers' measurements for each numbered part (for example, the range of observer A and B measurements for part 1, and so on) as shown in Table 8.3.

Step 3. Find the average range (in this case, the average range is .00022).

Table 8.3. Raw data—short form example.

Part number	Observer A	Observer B	Range
1	.5023	.5020	.0003
2	.5020	.5022	.0002
3	.5019	.5021	.0002
4	.5021	.5024	.0003
5	.5024	.5025	.0001

Step 4. Calculate the R&R error (RR). Note: The constant 4.33 value is equal to 99 percent of the area under the normal curve.

$$\text{RR} = \bar{R} \times 4.33 = .00022 \times 4.33 = .00095$$

Step 5. Compute the potential gage R&R%.

$$\text{GRR\%} = \frac{\text{ME}}{\text{Total tolerance}} \times 100 = \frac{.00095}{.005} \times 100 = 19\%$$

The measurement process is generally considered capable if the GRR% is 10 percent or less. It is considered marginal if the GRR% is between 10 percent and 25 percent. It is considered unacceptable if it exceeds 25 percent. These are general values. Individual companies must establish their own criteria (or meet required specifications for gage R&R).

ATTRIBUTE STUDIES

Attribute studies are used for measuring the overall effectiveness of observers inspecting attribute characteristics (for example, the results are good or bad). When working with attributes, the measures of repeatability and reproducibility do not apply as with variables. The following terms apply to attribute studies.

Uniformity is the overall agreement within and among observers.

Effectiveness (E) is the ability of the observers to detect both good and bad product.

Miss means accepting a bad part.

False alarm means rejecting a good part.

Bias is when observers are biased toward rejection or acceptance.

The primary purpose of attribute studies is to

1. Ensure that the acceptance criteria for the product and its characteristics have been properly established and communicated.
2. Identify and correct problems with the inspection acceptance criteria.
3. Identify observer-related problems and train observers accordingly.
4. Improve overall inspection effectiveness for product acceptance and/or process control purposes.

The Short-Form Attribute Study

Complete the following steps to perform a short-form attribute study.

Step 1. Identify a representative sample of parts for the study. This sample should be parts of the type and configuration that represent the purpose of the study. Number these parts 1 . . . *n* for future reference.

Step 2. Have those personnel who established the acceptance criteria (for example, engineering) inspect the sample and record their results. An example form is shown in Table 8.4.

Step 3. Have two or more observers inspect the sample of parts and those characteristics identified at least two times each, and record the results on a data sheet. Results are recorded as *good* or *bad*. Note: Have each observer inspect all *n* parts once, then all *n* parts again. Do not let the observers see each other's results.

In the example shown in Table 8.5, there are inconsistencies on the inspections of parts 3, 8, and 13. Part 3 and 13 shows inconsistent results with two inspections of Observer A

Table 8.4. Attribute study data sheet—short method.

Part	Observer A			Observer B		
	Trial 1	Trial 2	Trial 3	Trial 1	Trial 2	Trial 3
1						
2						
3						
4						
5						
6						
7						
8						
9						
10						
11						
12						
13						
14						
15						
16						
17						
18						
19						
20						

Table 8.5. An example of an attribute study.

Part	Observer A			Observer B		
	Trial 1	Trial 2	Trial 3	Trial 1	Trial 2	Trial 3
1	G	G		G	G	
2	G	G		G	G	
3	G	B		G	G	
4	B	B		B	B	
5	G	G		G	G	
6	G	G		G	G	
7	B	B		B	B	
8	B	B		G	G	
9	G	G		G	G	
10	G	G		G	G	
11	G	G		G	G	
12	G	G		G	G	
13	G	B		G	G	
14	G	G		G	G	
15	G	G		G	G	
16	G	G		G	G	
17	G	G		G	G	
18	G	G		G	G	
19	G	G		G	G	
20	G	G		G	G	

alone, and part 8 shows inconsistent results between observer A and B. The results of the short method is quite simply to identify inconsistencies within and between observer's results and take action to improve effectiveness.

The Long-Form Attribute Study

The long method for attribute studies requires the use of tables and calculations to complete the study and evaluate the results. It is a much more formal method. The following parameters must be used.

Effectiveness (E)

$$E = \frac{A + B}{\text{Total inspections}}$$

Probability of a false alarm (P_{Fa})

$$P_{Fa} = \frac{F}{A + F}$$

Probability of a miss (P_{Miss})

$$P_{Miss} = \frac{M}{B + M}$$

Bias (B) (Use Appendix C)

$$B = \frac{B_{Fa}}{B_{Miss}}$$

The quantity of good parts found correctly (A)

The quantity of bad parts found correctly (B)

Step 1. Refer to the purposes of attribute studies covered earlier in this chapter.

Step 2. Identify a representative sample of parts for the study. This sample should be parts of the type and configuration that represent the purpose of the study. Number these parts 1 . . . n for future reference.

Step 3. Have those personnel who established the acceptance criteria, inspect the sample and record their results.

Step 4. Have two or more observers inspect the sample of parts and those characteristics identified at least two times each, and record the results on a data sheet (shown in Table 8.6). Results are recorded as *good* or *bad*. Note: Have each observer inspect all n parts once, then all n parts again. Do not let the observers see each other's results.

Table 8.6. An example of a long method attribute study.

Total inspections = 84 (14 parts, 2 observers, 3 trials)

Part		Observer A			Observer B			Results			
		T1	T2	T3	T1	T2	T3	A	B	M	F
1	B	B	B	B	B	G	B	0	5	1	0
2	G	G	G	G	G	G	G	6	0	0	0
3	G	G	G	G	G	G	G	6	0	0	0
4	G	G	G	G	G	G	G	6	0	0	0
5	G	G	G	G	G	G	G	6	0	0	0
6	B	B	B	B	G	B	B	0	5	1	0
7	B	B	B	B	B	B	B	0	6	0	0
8	G	G	G	G	G	G	G	6	0	0	0
9	G	G	G	G	G	G	B	5	0	0	1
10	B	B	B	G	B	B	B	0	5	1	0
11	G	G	G	G	G	G	G	6	0	0	0
12	G	G	G	G	G	G	G	6	0	0	0
13	G	G	G	G	G	G	G	6	0	0	0
14	B	B	B	B	B	B	B	0	6	0	0
							Totals	53	27	3	1

As shown in Table 8.6, the following results were obtained.

Total inspections = 84

Total (A) = 53 (Good parts correctly accepted)

Total (B) = 27 (Bad parts correctly rejected)

Total (M) = 3 (Bad parts accepted)(misses)

Total (F) = 1 (Good parts rejected) (false alarms)

Based on the results in Table 8.6, the following parameters are calculated.

Effectiveness (E)

$$E = \frac{A + B}{\text{Total inspections}} + \frac{53 + 27}{84} = \frac{80}{84} = .952$$

Probability of a miss (P_{Miss})

$$P_{Miss} = \frac{M}{B + M} = \frac{3}{27 + 3} = \frac{3}{30} = 0.1$$

Probability of a false alarm (P_{Fa})

$$P_{Fa} = \frac{F}{A + F} = \frac{1}{53 + 1} = \frac{1}{54} = .02$$

Based on the probability of a false alarm (0.1) and the probability of a miss (0.02), the bias factors can be looked up in the table in Appendix C. The bias factors are $B_{Fa} = .0488$ and $B_{Miss} = .1758$. Bias, then, is calculated as follows:

Bias (B)

$$B = \frac{B_{Fa}}{B_{Miss}} = \frac{.0488}{.1758} = .28$$

Based on the computed parameters above, refer to Table 8.7 (acceptance parameters) to evaluate the results of the study. Using Table 8.7, the following results have been determined.

Effectiveness is acceptable (.952 is greater than 0.9).

P_{Fa} is acceptable (0.02 is less than 0.05).

Table 8.7. Acceptance parameters—attribute studies.

Parameter	Acceptable	Marginal	Unacceptable
Effectiveness	≥ .9	.8 to .9	< .8
P_{Fa}	< .05	.05 to .10	> 0.10
P_{Miss}	< .02	.02 to .05	> 0.05
Bias	.8–1.2	.5 to .8 or 1.2 to 1.5	< .5 or > 1.5

P_{Miss} is unacceptable (0.1 is greater than 0.05).

Bias is unacceptable (0.28 is less than 0.50).

Action should be taken, in general, as follows:

Improving the probability of misses (P_{Miss}) usually involves improving the acceptance criteria for the part (for example, clarifying the criteria, providing exacting samples of good versus bad, and so on).

Improving the bias (B) usually involves isolating the source of bias and training the observers accordingly.

REVIEW PROBLEMS

Refer to Appendix H for answers.

1. Which of the following elements are involved in the planning (or preparation phase) to perform a gage R&R study?
   a. Using calibrated measuring equipment
   b. Following the 10% rule of discrimination
   c. The proper gage has been selected
   d. All observers use the same (or similar) instrument
   e. All of the above
2. If the average of several measurements on a part is .5012, and the true value of that part is .501, what is the accuracy of the observer who made the measurements?
3. Regarding measurement error, the alpha risk is a problem for the producer? (True or false)
4. Parallax error falls into which of the following categories of measurement error?
   a. Manipulative
   b. Discrimination
   c. Observer
   d. Instrument
5. Getting consistent measurements repeatedly is related to the definition of (accuracy or precision).
6. What is the main reason why the measurements of one observer should not be seen by other observers involved in the gage R&R study?
   a. It's none of their business
   b. To avoid embarrassment
   c. To avoid bias
   d. None of the above
7. The typical gage R&R study only involves analysis of the repeatability and reproducibility of the measurement system. The added value of using standard parts of known value is the ability to measure ______________________.

8. Which of the following methods of gage R&R studies use constant K_1 and K_2 factors in the analysis? (short-range method or long-range method)
9. Of the two methods in question 8, which method is harder to perform in terms of calculations?
10. Which of the methods in question 8 is more accurate?
11. A gage R&R study has been completed and the total R&R percent is 37 percent. Is this generally acceptable?
12. Which of the following is one of the purposes of an attribute gage study?
    a. To blame people for poor judgment
    b. To speed up the inspection activity
    c. To improve overall inspection effectiveness
    d. None of the above
13. When performing an attribute study, what is the term used for the rejection of a good part?
14. Uniformity is the overall agreement within and among observers? (True or false)
15. With regard to personal judgment calls on the visual inspection and acceptance of products, which of the following will help reduce poor judgment calls?
    a. Preparing clearly defined acceptance criteria for quality
    b. Communicating the acceptance criteria to all observers
    c. Training observers as necessary
    d. Providing physical examples
    e. All of the above

CHAPTER 9

Regression and Correlation Analysis

There are times when quality control decisions have to be made that depend on knowledge about the relationship between two variables of a product or process. The variables, of course, depend on the product, process, or problem statement. There are many examples of regression and correlation applications in industry, and all of them are based on the question of whether two variables are correlated (one has a direct effect on the other). The primary tool used in regression and correlation analysis is the scatter diagram.

SCATTER DIAGRAM APPLICATIONS

Some examples of industrial applications of regression and correlation analysis follow.

Testing—Two different tests are evaluated to see if they show correlated results. The purpose of the correlation study might be to help make decisions about which test to use. If there is a high correlation between the test results, one might choose the less expensive test. Since the less expensive test correlates to the other, testing costs are reduced.

Process control—Two process parameters are being considered for control chart purposes. If the variables shows high correlation to the expected process output (or result), the process parameter, if controlled, can improve the output. An example of this could be to study the correlation between the temperature of a heat treat oven and the hardness of the products. Other examples could be to test the relationship between speed and/or feed rates to the surface finish of a machined part or to test the relationship between the time parts spend in a plating tank and the plating thickness of those parts. Note, however, these examples are only food for thought. They do not necessarily correlate or there may be other variables interacting on the output. If this were so, the scatter diagram may not be the correct tool. The situation may call for multivariate experiments.

Measuring and gaging—Similar to the testing example, where two different measurement systems are evaluated to see if they correlate. If so, sound decisions could be made about which measurement system to use.

Skills versus output quality—A group of different operators run an operator-dominant process (one in which the quality of the output is assumed to depend upon the knowledge, skill, and/or dexterity of the operator). A scatter diagram can help test the strength of the relationship between numerical indexes that define their skills and the output variable of concern. In this case, the output variable is the dependent (Y) variable, and the knowledge skill index is the independent (X) variable.

DATA COLLECTION FOR SCATTER DIAGRAMS

When using scatter diagrams, the variable that may (or may not) be affected is referred to as the *dependent variable* (Y). The variable that may have an affect on the dependent variable is called the *independent variable* (X). For example, if it is assumed that the temperature of a heat treat oven has a direct affect on part hardness, the oven temperature is the independent variable (X) and hardness is the dependent variable (Y).

The examples to follow assume that there was reason to test the relationship between two variables (X and Y) in a process. The data collected for a scatter diagram must be in paired sets. Paired sets of data means that a given measurement of the independent variable (X) is paired with the corresponding measurement made on the dependent variable (Y). In other words, the Y variable measurement recorded on the part must be associated directly with the X variable measurement of that part.

Another example is if n parts were tested with two different tests (X and Y), the results of test X and test Y for each part would be recorded side-by-side (paired sets). Consider the data shown in Table 9.1. Notice how the values of X and Y for each part are recorded in paired sets.

PLOTTING THE SCATTER DIAGRAM

The graph of a scatter diagram has an X (horizontal) and Y (vertical) scale that fits all the data collected. Scatter diagrams are plotted using plot points that represent each paired set of data. For example, if the X and Y (X, Y) values for one part are 5, 16, respectively, a point would be plotted on the diagram that represents an X magnitude of 5 (on the X-axis scale) and a Y magnitude of 16 (on the Y axis scale). Typically, if there is more than one plot point with the same X, Y values, then a point is used first and other identical plot points are represented by a circle around the point. If, for example, there were three parts that had the

Table 9.1. Example of data for positive correlation in paired sets.

X	Y	X^2	Y^2	XY
2	13	4	169	26
7	21	49	441	147
9	23	81	529	207
1	14	1	196	14
5	15	25	225	75
12	21	144	441	252

same X, Y results, there would be a plot point on the scatter diagram and two circles around that point. See Figure 9.1 for an example of points that are circled because of more than one set of the same X, Y values. Once all of the paired sets of data are plotted, one can begin to interpret the results.

CORRELATION AND INTERPRETATION

Correlation means that a change in the independent variable (X) causes a proportionate change in the dependent variable (Y). *Positive correlation* means that an increase in the X variable causes an increase in the Y variable. *Negative correlation* means that an increase in the X variable causes a decrease in the Y variable. *No correlation* means that there is no relationship between the two variables, so one cannot expect specific changes in Y for specific changes in X.

Estimating Y *Values for Given* X *Values*

When there is correlation between two variables X and Y, then for any given X value, a Y value can be calculated. There are two methods for computing Y for any given X value: visual identification and calculation.

There are times when the formal regression and correlation statistics are not necessary because the relationship (or lack of relationship) between two variables is clearly shown by the scatter diagram. The pattern of plotted points on the scatter diagram indicates if two variables are correlated (related) or not. The regression and correlation statistics add numerical value to the information. This is not to say that the statistics should not be computed. It simply means that in this case, the "picture is worth a thousand words."

When there is positive correlation, the cluster of plotted points will appear at a specific slope rising from left to right on the scatter diagram. Refer to Figure 9.2 for an example of positive correlation. When there is negative correlation, the cluster of plotted points will appear at a specific slope declining from left to right on the scatter diagram. Refer to Figure 9.4 for an example of no correlation. Refer to Figure 9.6 for an example of negative correlation. When there is no correlation, the plotted points will not cluster tightly (for example, all over the graph) and they will have no definite slope.

As shown in Figure 9.9, the Y value could have been estimated visually be finding 5 on the X scale, looking straight up from that point to the regression line, then looking straight to the left at the Y scale (which is about 17).

The other method of computing X and Y values is by calculation. In Figure 9.3, the software program has compiled a Y value of 16.9356 for an entered value of 5. For the example of computing Y for any given X value, the operator-dominant process (Table 9.1) data will be used. First, the a and b values must be computed using the following equations.

$$a = \frac{(\Sigma y)(\Sigma x^2) - (\Sigma x)(\Sigma xy)}{n(\Sigma x^2) - (\Sigma x)^2} = \frac{(107)(304) - (36)(721)}{6(304) - (36)^2} = 12.447$$

$$b = \frac{n(\Sigma xy) - (\Sigma x)(\Sigma y)}{n(\Sigma x^2) - (\Sigma x)^2} = \frac{6(721) - (36)(107)}{6(304) - (36)^2} = .898$$

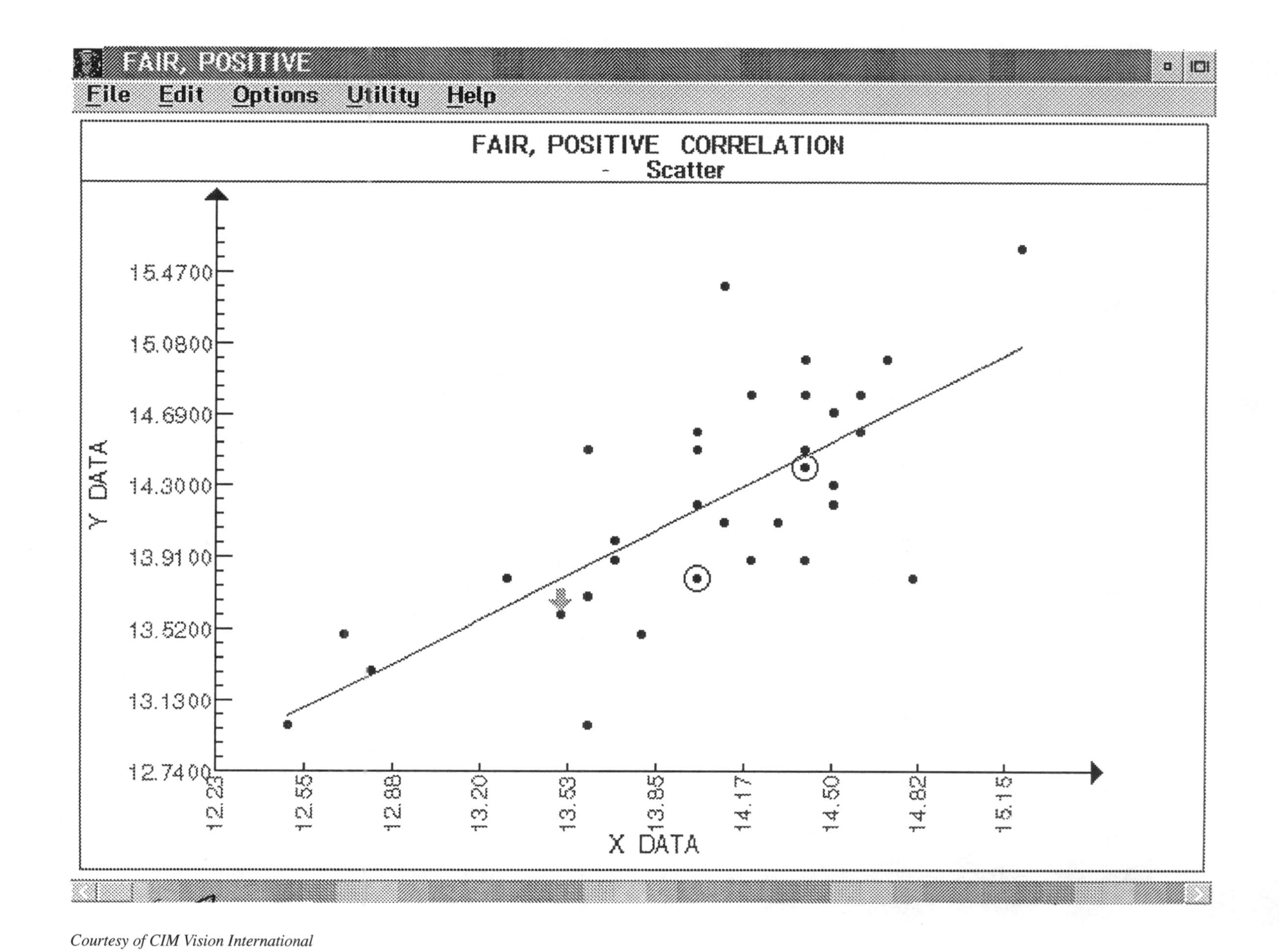

Courtesy of CIM Vision International

Figure 9.1. Scatter diagram showing fair, positive correlation. Duplicate plot points are circled.

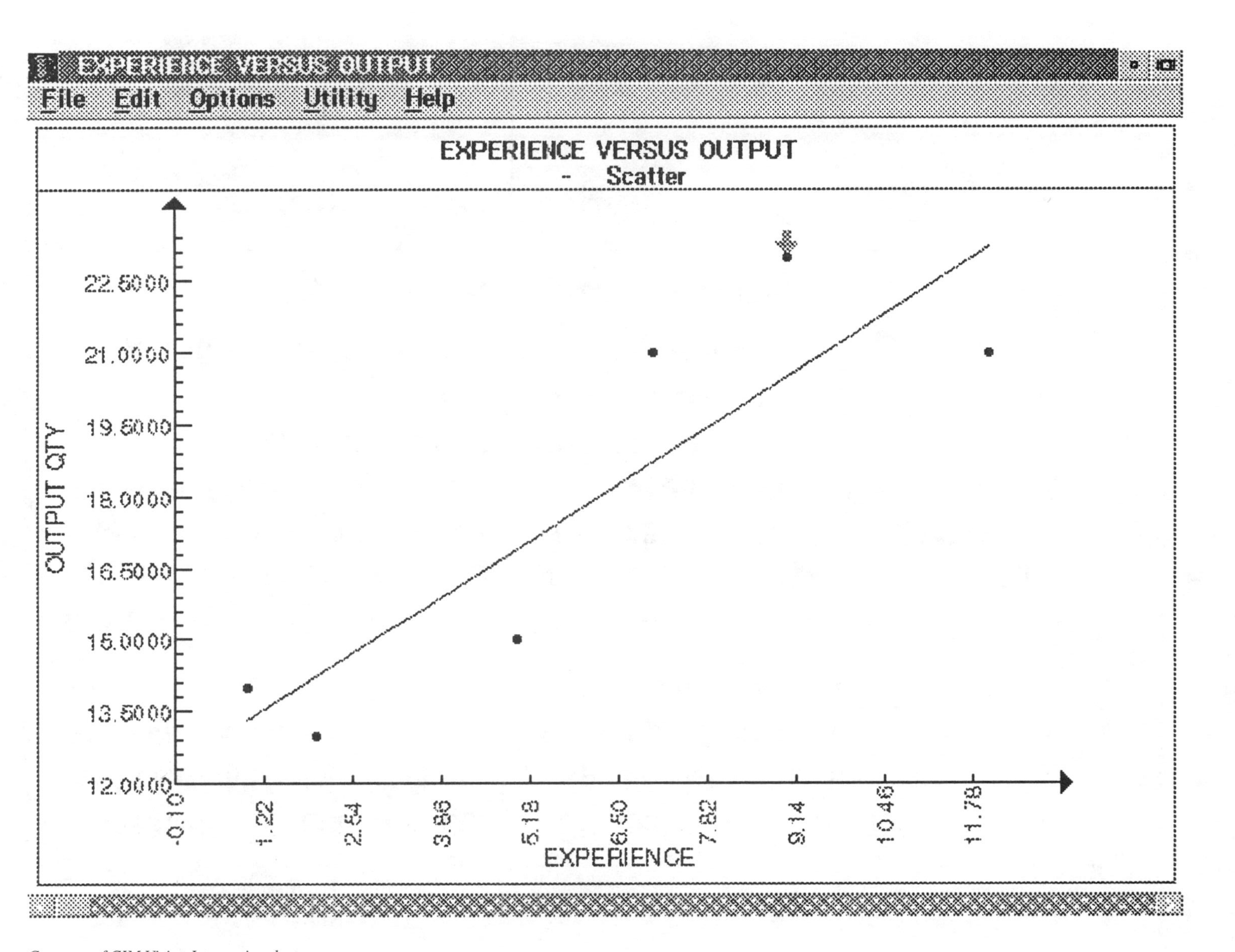

Courtesy of CIM Vision International

Figure 9.2. Scatter diagram showing positive correlation.

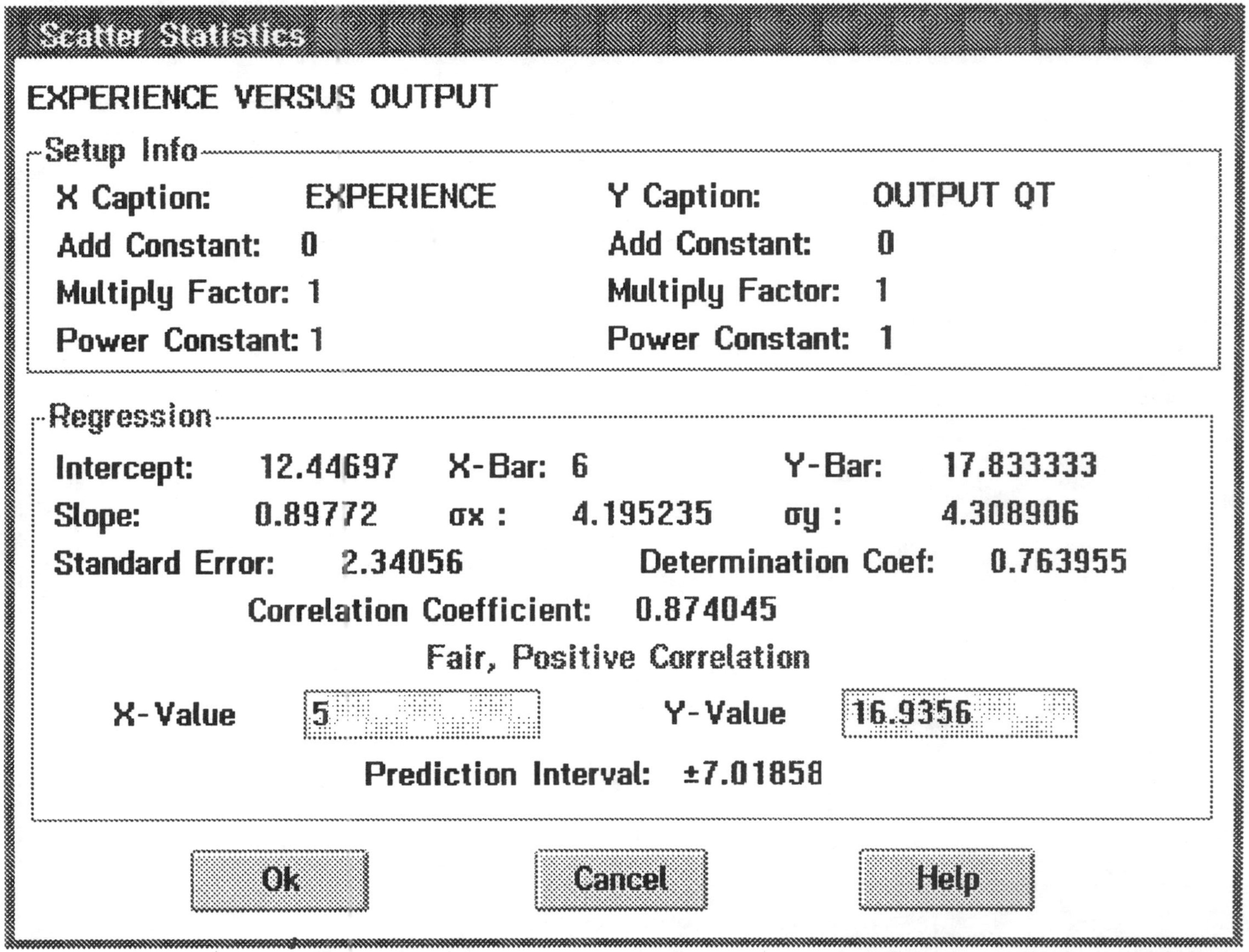

Scatter Statistics

EXPERIENCE VERSUS OUTPUT

Setup Info

X Caption:	EXPERIENCE	Y Caption:	OUTPUT QT
Add Constant:	0	Add Constant:	0
Multiply Factor:	1	Multiply Factor:	1
Power Constant:	1	Power Constant:	1

Regression

Intercept:	12.44697	X-Bar:	6	Y-Bar:	17.833333
Slope:	0.89772	σx :	4.195235	σy :	4.308906

Standard Error: 2.34056 Determination Coef: 0.763955

Correlation Coefficient: 0.874045

Fair, Positive Correlation

X-Value 5 Y-Value 16.9356

Prediction Interval: ±7.01858

Ok Cancel Help

Courtesy of CIM Vision International

Figure 9.3. Regression and correlation statistics for a positive correlation.

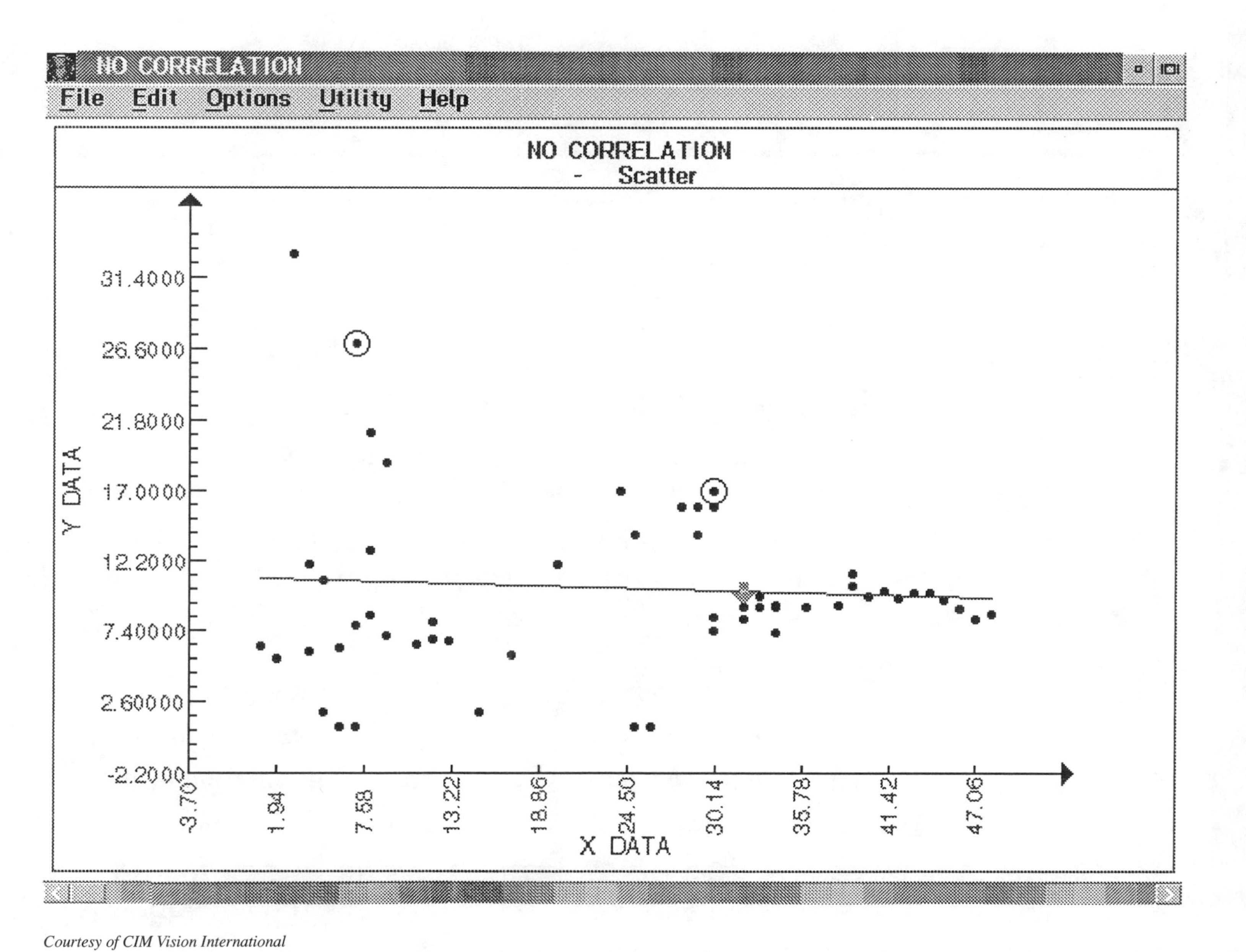

Courtesy of CIM Vision International

Figure 9.4. Scatter diagram showing no correlation.

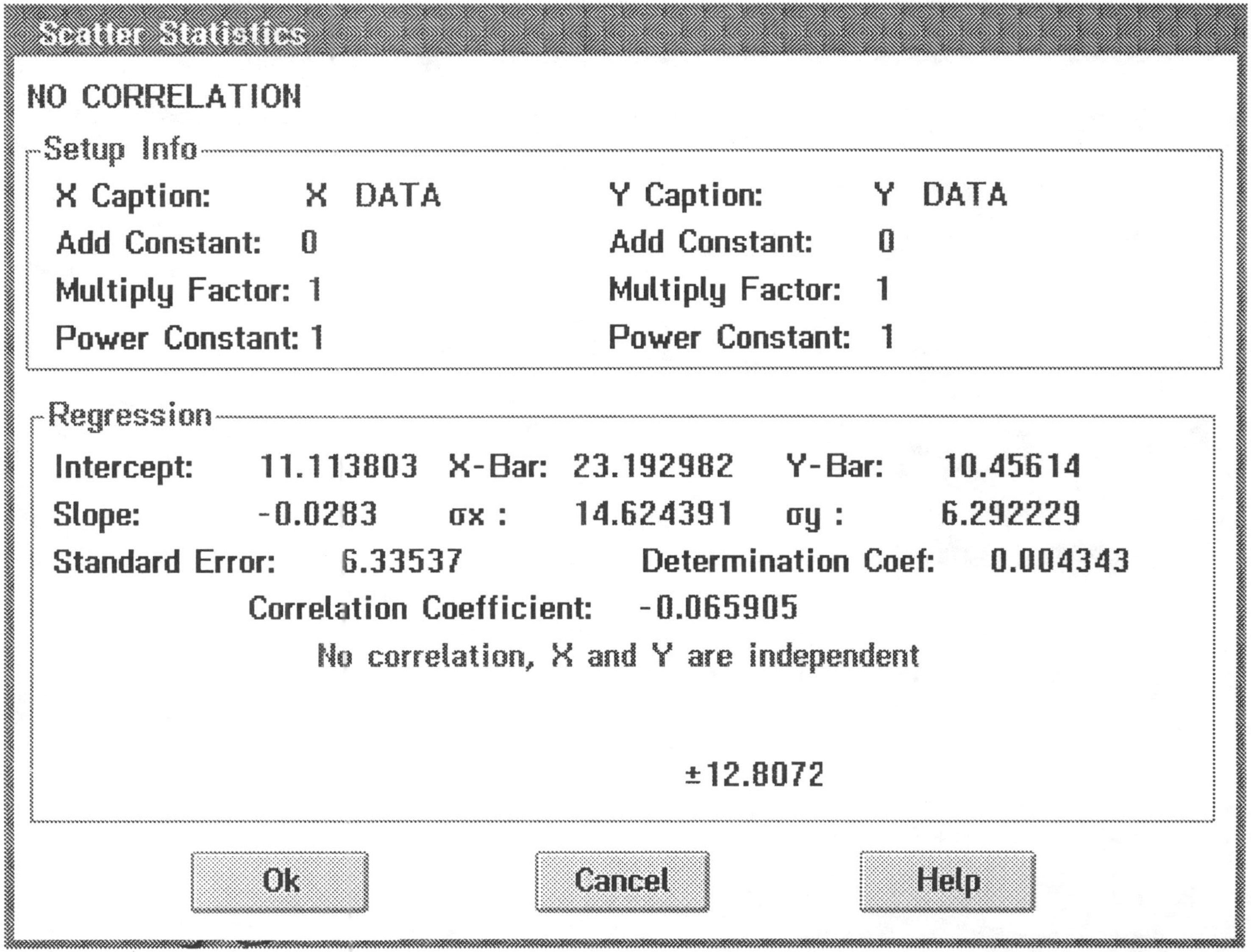

Scatter Statistics

NO CORRELATION

Setup Info

X Caption:	X DATA	Y Caption:	Y DATA
Add Constant:	0	Add Constant:	0
Multiply Factor:	1	Multiply Factor:	1
Power Constant:	1	Power Constant:	1

Regression

Intercept:	11.113803	X-Bar:	23.192982	Y-Bar:	10.45614
Slope:	-0.0283	σx :	14.624391	σy :	6.292229

Standard Error: 6.33537 Determination Coef: 0.004343

Correlation Coefficient: -0.065905

No correlation, X and Y are independent

±12.8072

Ok Cancel Help

Courtesy of CIM Vision International

Figure 9.5. Regression and correlation statistics for no correlation.

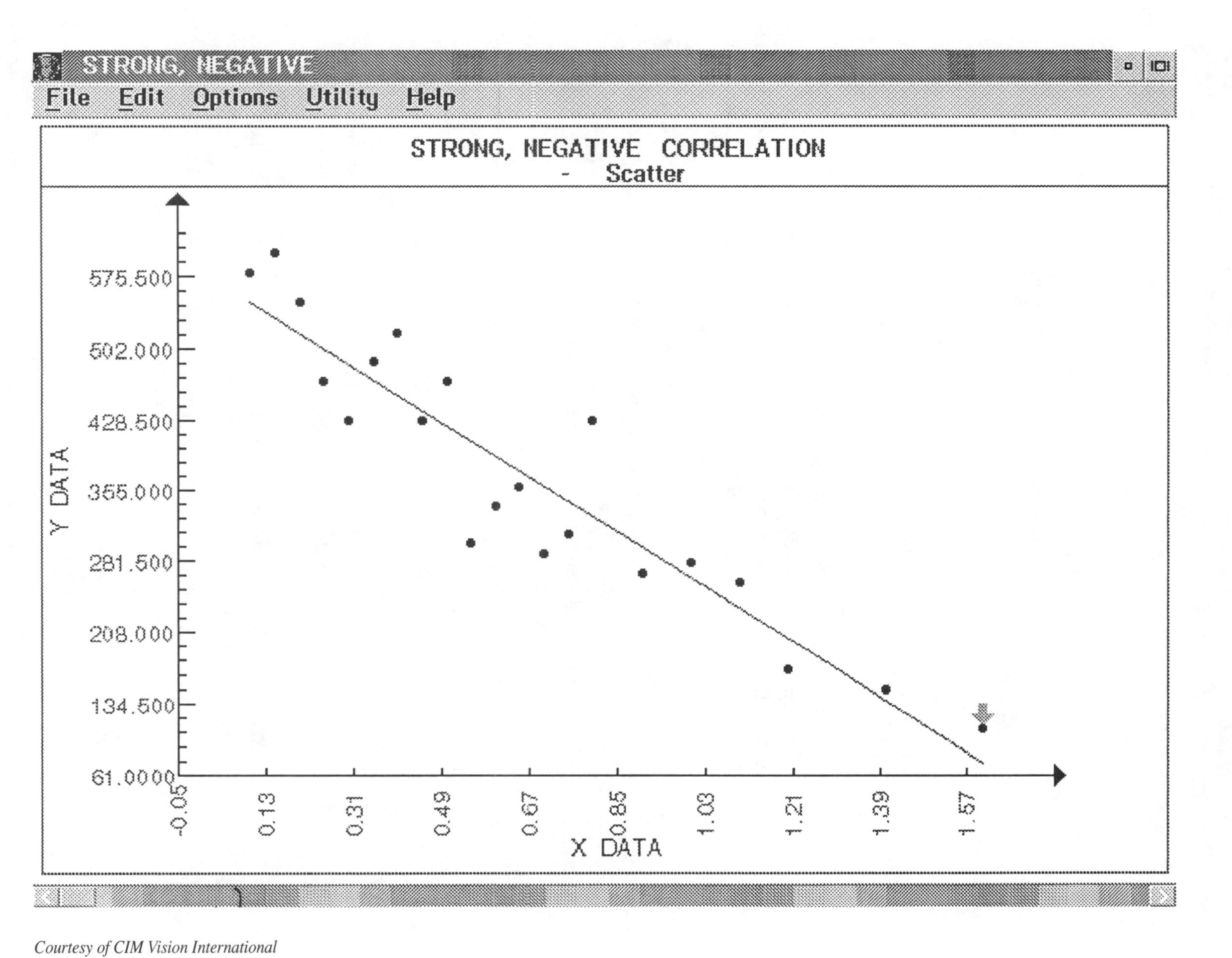

Courtesy of CIM Vision International

Figure 9.6. Scatter diagram showing negative correlation.

A resulting Y value for any given X value is then computed using the following equation.

$$Y = a + bx = 12.447 + (.898 \bullet 5) = 17$$

where

x is the given X value (5 years, in this case)

In this case, a person with 5 years experience can be expected to have an output of about 17 units. Note that, if a scatter diagram were plotted, one could have visually estimated the Y value of 17 units for the given X value of 5 years experience.

Regression and Correlation Statistics

It is not the purpose of this chapter to cover all regression and correlation statistics in depth, but the following statistics are pertinent to the scatter plot in any application. Figures 9.3, 9.5, 9.7, and 9.8 show regression and correlation statistics for each of the examples. The following in an explanation of the most pertinent statistics.

Correlation coefficient is a measure of the strength of the relationship. A coefficient of 1.0, for example, indicates perfect positive correlation, where a coefficient of –1.0 indicates perfect negative correlation.

$\bar{X}$ represents the average of the values for the X variable.

$\bar{Y}$ represents the average of the values for the Y variable.

σ_X represents the standard deviation of the X values.

σ_Y represents the standard deviation of the Y values.

Determination coefficient is the square of the correlation coefficient.

Intercept is the point at which the regression line intersects the Y axis.

Slope is the rate of change along the regression line. The slope is the ratio of the vertical distance to the horizontal distance (rise versus run) between any two points on the regression line.

Standard error is a measure of the variability around the regression line. For example, the variability of predicted Y values for given X values.

More detailed information about these correlation and regression statistics is beyond the scope of this chapter. Refer to the bibliography for more references.

Calculating the Correlation Coefficient

Calculating the correlation coefficient is an involved equation, but if it is broken down into its parts, it is much easier. Consider the following example for calculating the correlation coefficient is an operator-dominant process where X is the years of experience of the operators and Y is the output quantity. Table 9.1 shows the X, Y data for the problem. The table also includes columns that help find the other products required in the equation (such as the values of X^2).

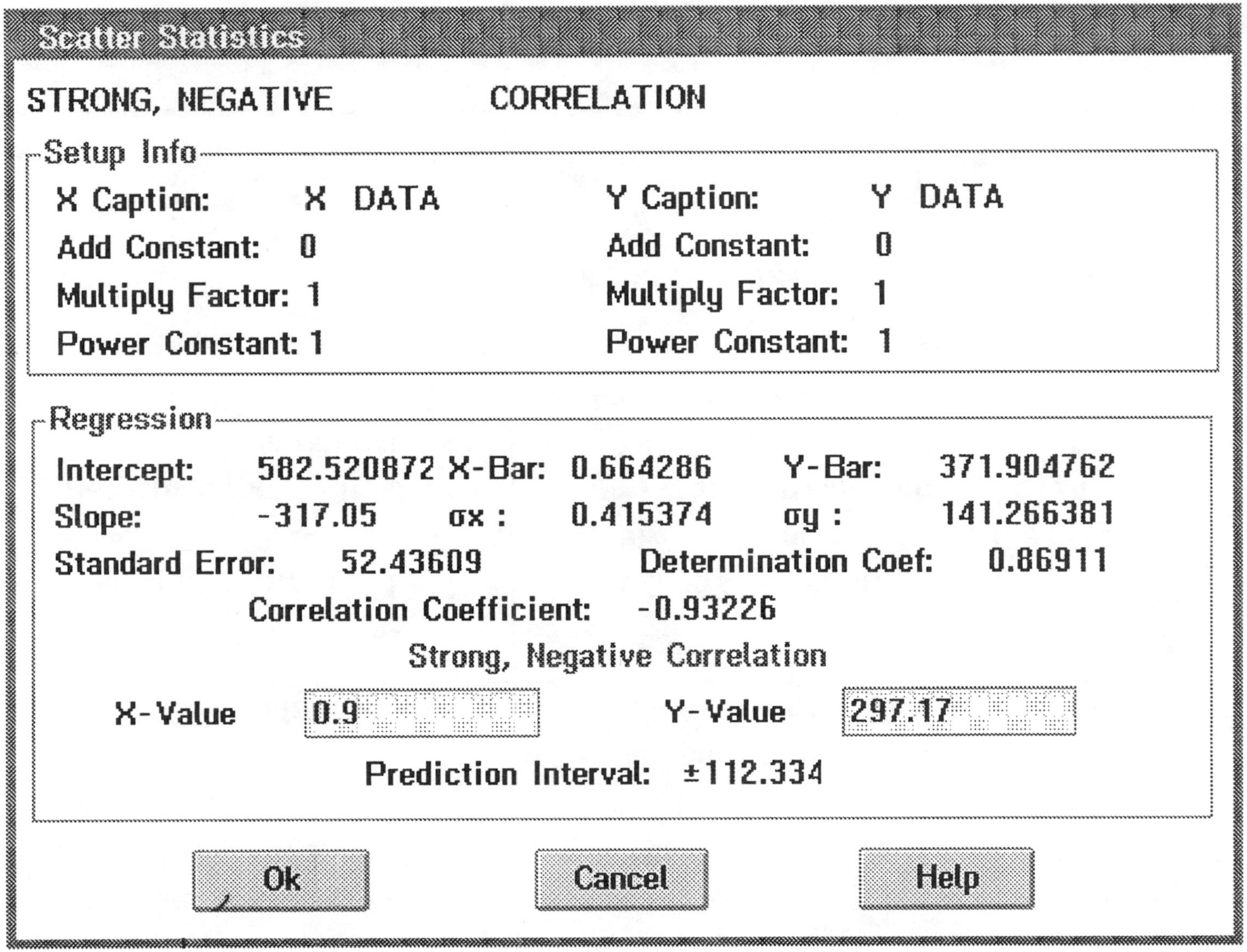

Scatter Statistics

STRONG, NEGATIVE CORRELATION

Setup Info

X Caption:	X DATA	Y Caption:	Y DATA
Add Constant:	0	Add Constant:	0
Multiply Factor:	1	Multiply Factor:	1
Power Constant:	1	Power Constant:	1

Regression

Intercept:	582.520872	X-Bar:	0.664286	Y-Bar:	371.904762
Slope:	-317.05	σx :	0.415374	σy :	141.266381

Standard Error: 52.43609 Determination Coef: 0.86911

Correlation Coefficient: -0.93226

Strong, Negative Correlation

X-Value 0.9 Y-Value 297.17

Prediction Interval: ±112.334

Ok Cancel Help

Courtesy of CIM Vision International

Figure 9.7. Regression and correlation statistics for a strong negative correlation.

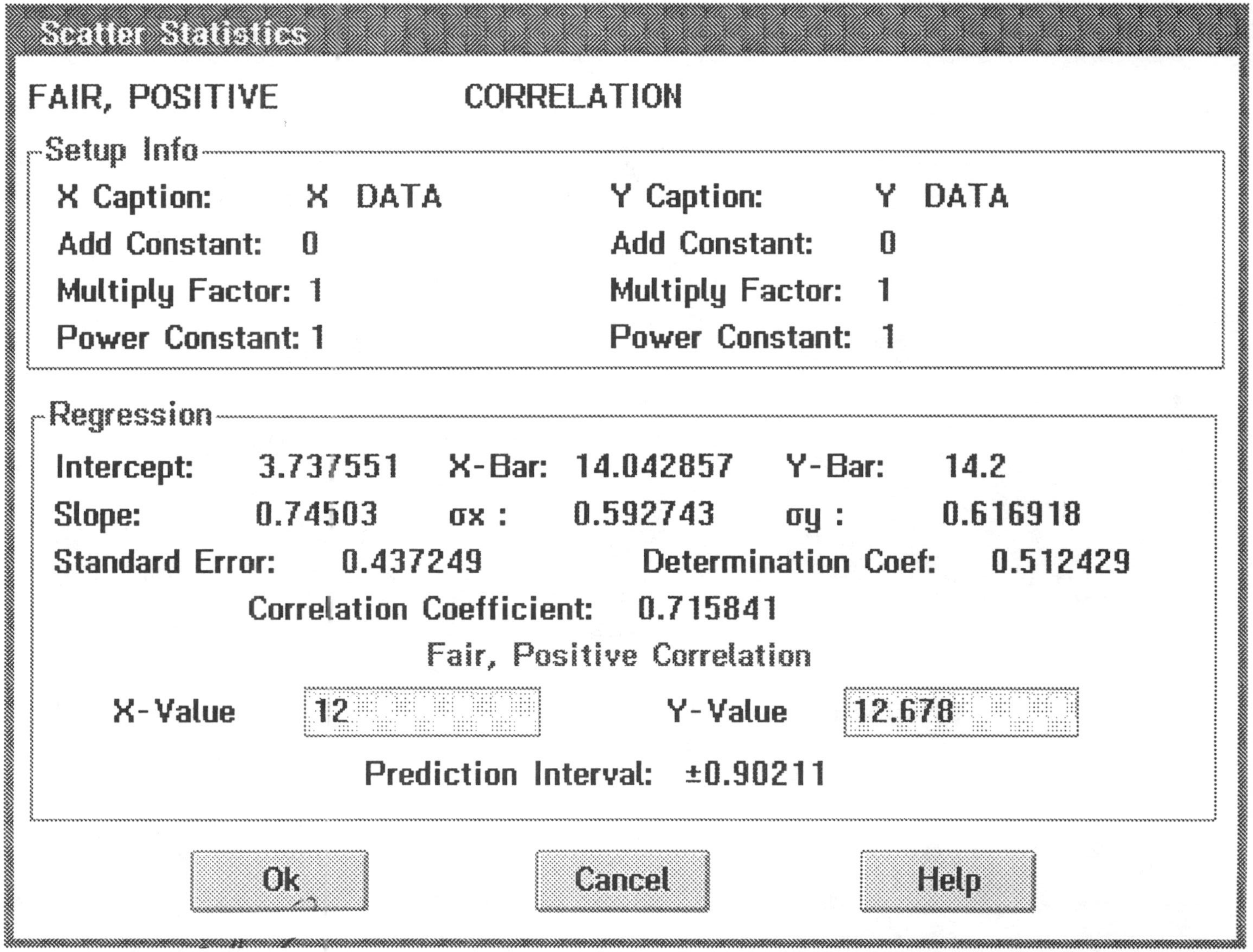

Scatter Statistics

FAIR, POSITIVE CORRELATION

Setup Info

X Caption:	X DATA	Y Caption:	Y DATA
Add Constant:	0	Add Constant:	0
Multiply Factor:	1	Multiply Factor:	1
Power Constant:	1	Power Constant:	1

Regression

Intercept:	3.737551	X-Bar:	14.042857	Y-Bar:	14.2
Slope:	0.74503	σx :	0.592743	σy :	0.616918

Standard Error: 0.437249 Determination Coef: 0.512429

Correlation Coefficient: 0.715841

Fair, Positive Correlation

X-Value 12 Y-Value 12.678

Prediction Interval: ±0.90211

Ok Cancel Help

Courtesy of CIM Vision International

Figure 9.8. Regression and correlation statistics for a fair and positive correlation.

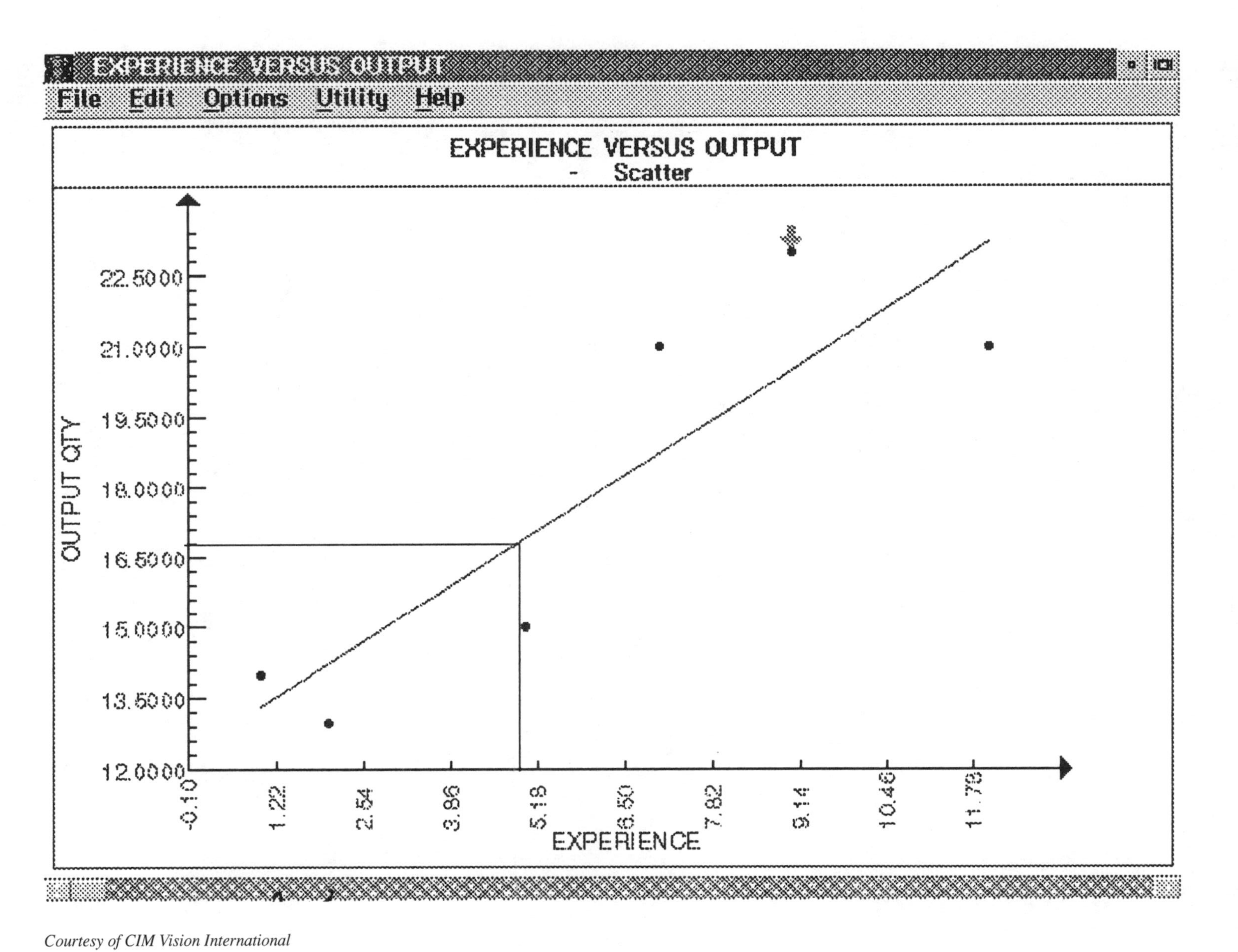

Courtesy of CIM Vision International

Figure 9.9. Visually estimating *Y* values for given *X* values.

The correlation coefficient equation is as follows:

$$\tau = \frac{n(\Sigma xy) - (\Sigma x)(\Sigma y)}{\sqrt{n(\Sigma x^2) - (\Sigma x)^2} \bullet \sqrt{n(\Sigma y^2) - (\Sigma y)^2}}$$

$$\tau = \frac{6(721) - (36)(107)}{\sqrt{6(304) - (36)^2} \bullet \sqrt{6(2001) - (107)^2}} = .874$$

The answer for the correlation coefficient (.874) indicates fair positive correlation between the years of experience and the output quantity of the operators. These results are shown in Figure 9.2.

Interpreting the Correlation Coefficient

In general, correlation coefficients can be interpreted where the results indicate strong, fair, weak, or no correlation. The criteria in Table 9.2 provide useful guidelines for interpreting correlation.

Table 9.2. Criteria for interpreting correlation coefficient.

Correlation	Range of coefficients
Strong, positive	≥ 0.9 to 1.0
Fair, positive	≥ 0.7 to < 0.9
Weak, positive	≥ 0.5 to < 0.7
No correlation	Between –0.5 and +0.5
Weak, negative	≥ –0.5 to < –0.7
Fair, negative	≥ –0.7 to < –0.9
Strong, negative	≥ –0.9 to –1.0

REVIEW PROBLEMS

Refer to Appendix H for solutions.

1. Which of the two variables under study (*X* or *Y*) is the independent variable?
2. A scatter diagram shows strong positive correlation. This means that when the *X* variable increases, the *Y* variable (increases or decreases)?
3. When both *X* and *Y* variables are measured on each part and recorded side-by-side, the data are called ____________________ data.
4. The point along the *Y* axis of a scatter diagram where the regression line intersects is called the ____________________.
5. The correlation ____________________ is the value that numerically describes the strength of the relationship between variables *X* and *Y* (if there is a relationship).

6. When there is correlation, using the scales of the X and Y axes, and the regression line, one can visually estimate a Y value for any given X value. (True or false)
7. Which of the following purposes does not involve the use of a scatter diagram for improvement?
   a. To test the relationship between two variables
   b. To evaluate two different measurement methods
   c. To evaluate six variables and their interaction
   d. To evaluate two candidate control parameters
8. Two variables X and Y are correlated with a correlation coefficient of –.98. In this case, an increase in the X variable will cause a ____________________ in the Y variable.
9. If a scatter diagram is being plotted and there are two sets of data that have the same plot point value, a plot point and a ____________________ will be used to plot both sets.
10. A wide cluster of plotted points on a scatter diagram indicates that correlation (does/does not) exist between variables X and Y.

CHAPTER 10

Problem-Solving Techniques and Tools

INTRODUCTION

The following five basic problem-solving steps have been proven to be effective. These steps, in order, are

1. Identify the problem.
2. Identify the root or true cause.
3. Take action to correct or eliminate the cause.
4. Follow up on the action taken.
5. Monitor or control the level achieved.

In general, these steps work because they offer a logical progression toward the solution of the problem. For example, it is difficult to identify the cause of a problem without a clear understanding of the problem itself. There are also a variety of tools that can be used during the various steps of problem solving. These tools are

- Brainstorming
- Cause-and-effect diagrams
- Pareto analysis
- Check sheets
- Matrix diagrams
- Run charts
- Control charts
- Scatter diagrams
- Histograms

The following covers each problem-solving step with a brief discussion and explanation of how it works. Also included are some examples of the problem-solving tools that might be used.

THE FIVE PROBLEM-SOLVING STEPS AND TOOLS

Identify the Problem

Problem Statement. It is important to establish a clear problem statement during the first phase of problem solving. Problem statements that are too general or ambiguous can cause delays, wasted time, and, more importantly, ineffective solutions.

An everyday example of a problem statement is "car won't start." If this is the clearest statement available, then continue. The subsequent problem-solving activity, however, could take several cause identification paths, which would confuse the issue. A better problem statement, if it is true, may be "starter won't turn over." This clearer problem statement narrows down the probable causes significantly and enhances the following steps for an effective solution.

1. Take the time to make sure that problem statements are clearly understood by everyone involved. Brainstorming can help in this case.
2. Describe or narrow the problem statement as much as possible.
3. Make sure there is some form of standard available for comparison. Resolve unclear or missing standards before proceeding with problem solving.

Problem Description. Describe the problem in as much detail as necessary in order to define the degree of the problem, its impact on other areas or operations, and those individuals who are best equipped to solve the problem.

Problem Severity. Attempt to define the severity or priority of problems using data. Data can be found in reports, check sheets, computer files, or other sources. Data can then be transformed into information using various graphs, control charts, statistical analysis, Pareto analysis, and other methods.

Identify the Root or True Cause

Cause Versus Symptom. Identifying the root cause of a problem is the next important step. Be careful to look for problem causes, not symptoms. An everyday example of a symptom is the cough that usually accompanies a common cold. Coughing is a symptom of the cold, but not the cause. The cause could be several things such as walking barefoot or low body resistance to viruses.

All problems have one or more true causes that are usually difficult to find and one or more symptoms that are often easy to find. Taking action to correct a symptom will rarely solve the problem and may even cover up the cause. Cough syrup will not cure the common cold nor will it prevent the cold from returning.

Cause-and-Effect (Fishbone) Diagrams. Cause-and-effect diagrams are a helpful problem-solving tool that is used to analyze a problem based on six possible cause categories. These are

- Manpower
- Materials
- Machines

- Methods
- Measurement
- Environment

More often than not, the true cause(s) of a problem can be found in one or more of these categories. The cause-and-effect diagram has been nicknamed the fishbone diagram because of its similarity to the spine and ribs of a fish, as shown in Figure 10.1. The problem statement (or effect) is written at the far right side and the six ribs or areas are shown leading to the problem.

Cause-and-effect diagrams provide a method for stimulating thoughts in an organized manner so that possible causes can be identified and expanded upon by the problem-solving team. Once this diagram is laid out, the problem-solving team can brainstorm possible causes for further investigation.

Figure 10.2 is a brief example of the previous problem statement "car won't start" in which the cause-and-effect diagram and the brainstorming are completed. Note that this is only a brief example, and also note that the clearer problem statement "starter won't turn over" eliminates many of the causes identified.

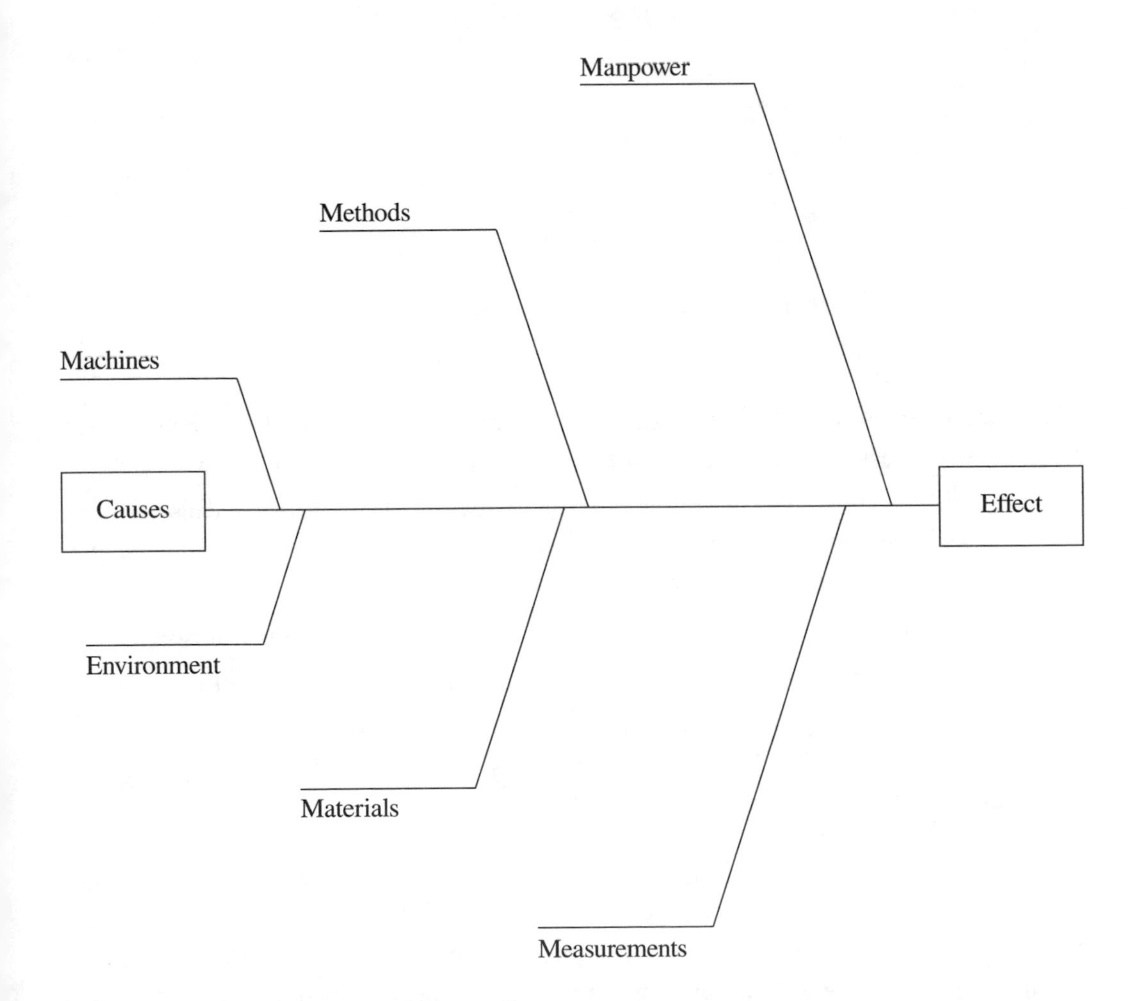

Figure 10.1. Cause-and-effect (fishbone) diagram.

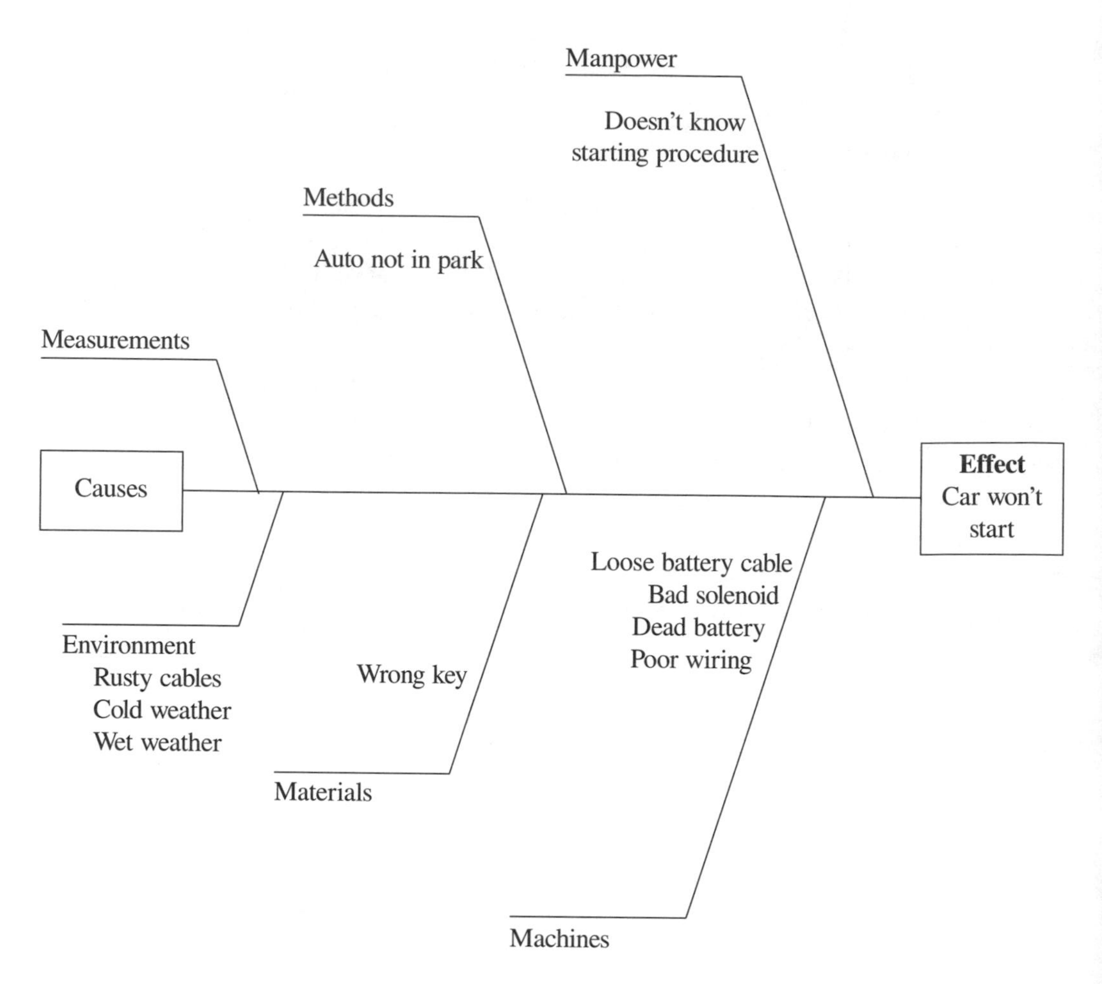

Figure 10.2. Cause-and-effect diagram for "car won't start."

Brainstorming. Brainstorming is used to identify all possible causes of the problem. The basic steps to structured brainstorming are

1. Each individual on the team takes a turn at identifying one possible cause no matter how wild the idea may seem. The ideas are not laughed at or discussed, just listed. The team member may pass if an idea does not come to mind.
2. The initial brainstorming is over when all participants pass in consecutive turns.
3. Each idea is then highlighted, discussed, voted upon, and then kept or eliminated.
4. Ideas that receive the highest scores in the voting process are further investigated. Upon investigation, the ideas are verified or eliminated. This brainstorming and elimination process often identifies the true cause(s) of the problem.

Pareto Analysis (Several Causes). In cases where several causes for a problem are present, Pareto analysis (with data) can be used to arrange the causes in order to locate the largest contributor so that it can be acted upon first. Using Pareto analysis, a problem-solving team can make a significant improvement in a short time with the least amount of effort. Pareto analysis requires that attribute or variable data are collected, then analyzed to separate the vital few causes versus the trivial many. Check sheets (Figure 10.3) can be used for consistent data collection (variables or attributes). Most problems that have multiple causes can then be analyzed to find the few causes at the root of the problem. When these few causes are corrected, the impact of the problem is reduced and the other causes can then be corrected for final solution. Consider the Pareto analysis in Figure 10.4.

In Figure 10.4, it is obvious that if the undersize O.D. problem is solved first, the severity of the overall scrap problem will be reduced by over 50 percent. It would be a shame to concentrate problem-solving efforts on the trivial many causes, such as cracks and nicks, and continue to scrap over 50 percent of the shafts produced because of the undersize condition.

Multiple Pareto Analyses. Pareto analysis is a powerful tool for problem solving, but, at times, one Pareto analysis can be misleading. As mentioned in chapter 4, I never use just

Shaft line defect	Quantity scrap	Total	Percent of total
Cracks	XXX	3	5%
Bent	XXXXXXXXXXXXXXX	15	25%
O.D. U/S	XXXXXXXXXXXXXXXXXXXXXXXXXXXXXX	30	50%
Tool marks	XXX	3	5%
Nicks	XXX	3	5%
Miscellaneous	XXXXXX	6	10%

Figure 10.3. Check sheet for data collection (Pareto style).

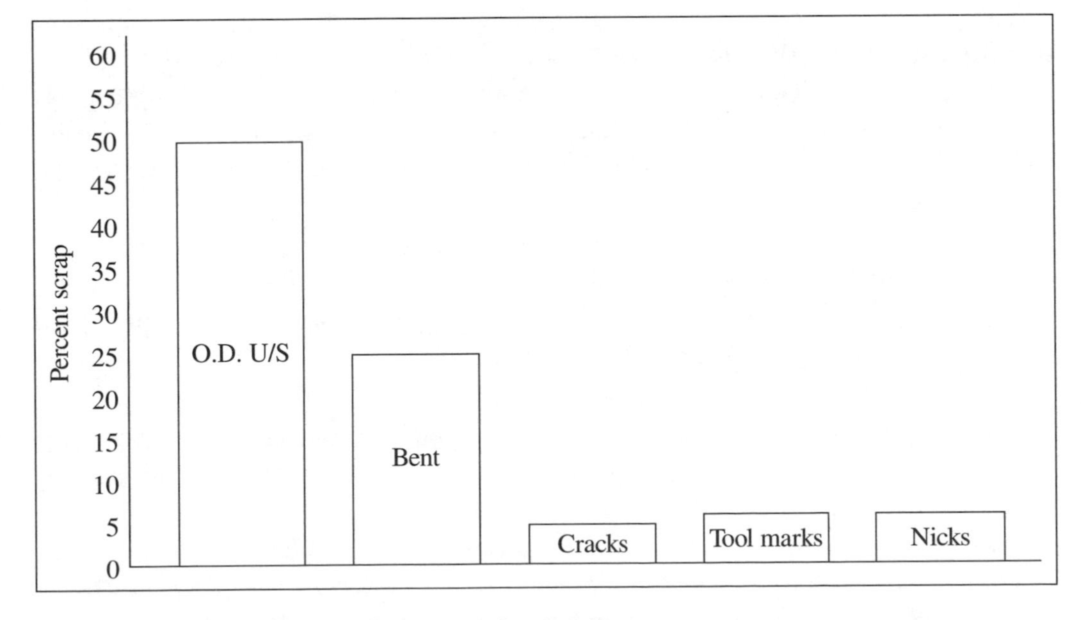

Figure 10.4. Pareto analysis (percent scrap) for shaft line.

one Pareto analysis. For example, refer to the Pareto analysis on housing casting defects in Figure 10.5. The Pareto analysis in Figure 10.5 is a frequency analysis that shows the top contributors by their frequency (or quantity). It shows that the defect porosity represents about 54 percent of the total quantity of defects for the housing castings. It also shows a cumulative line that indicates the top two defects (porosity and excess grind) represent about 74 percent of all defects. The rest of the defects represent the remaining 26 percent of the defects for the housing castings. Looking at this Pareto analysis, one might tend to begin working on reducing porosity and excess grind.

There are other ways to perform Pareto analysis (other than by the frequency of defects). For example, consider that not all defects always have the same level of importance from a product or cost standpoint. At times, defects have either been classified (such as minor, major, or critical) or they have been assigned weights in accordance with their importance. Another Pareto analysis for the same defects has been prepared in Figure 10.6. Figure 10.6 is a weighted Pareto analysis chart, in which each defect has been assigned a specific weight (on a scale from 1 to 100) and the analysis first multiplies the frequency of each defect times its weight before plotting the Pareto chart.

Figure 10.6 could also be a cost analysis in which the weight of the defect reflects the cost associated with the defect. Recall that the frequency Pareto chart in Figure 10.5 showed that porosity and excess grind were the top two contributors. The weighted chart in Figure 10.6 now shows an entirely different picture. Core shift and cold shut have taken over first and second place from a weight (or cost) standpoint.

Performing more than one Pareto analysis can give the problem-solving team considerably more information upon which to make a sound decision on severity of problems.

Take Action to Correct or Eliminate the Root Cause

Once the root cause has been identified for a problem, specific action should be taken to eliminate the cause and thereby correct the problem.

Identify the Group Responsible. The root cause of a problem is the key to identifying the specific group within the company that is responsible for action. Often, this group has already been identified in an earlier problem-solving step (see *problem description*), however, other groups may share responsibility for action when the cause crosses departmental lines of responsibility.

The best approach toward identifying the responsible group (or team) for problem solving is to look for the natural work team. A *natural work team* is a group of people who are closest to the problem (or who own the problem). In general, the natural work team to solve a problem is automatically picked by the problem statement.

Select the Specific Action. Once the responsible group is identified, the brainstorming technique can be used again to help select the action that will be taken. Often, matrix diagrams are useful in analyzing and comparing different alternatives for problem solution. The matrix can be used to show a variety of different results for different alternatives (see Figure 10.7), or it can be used to identify problems that are generic to different processes. The matrix is useful when there are several possible outcomes with several possible effects.

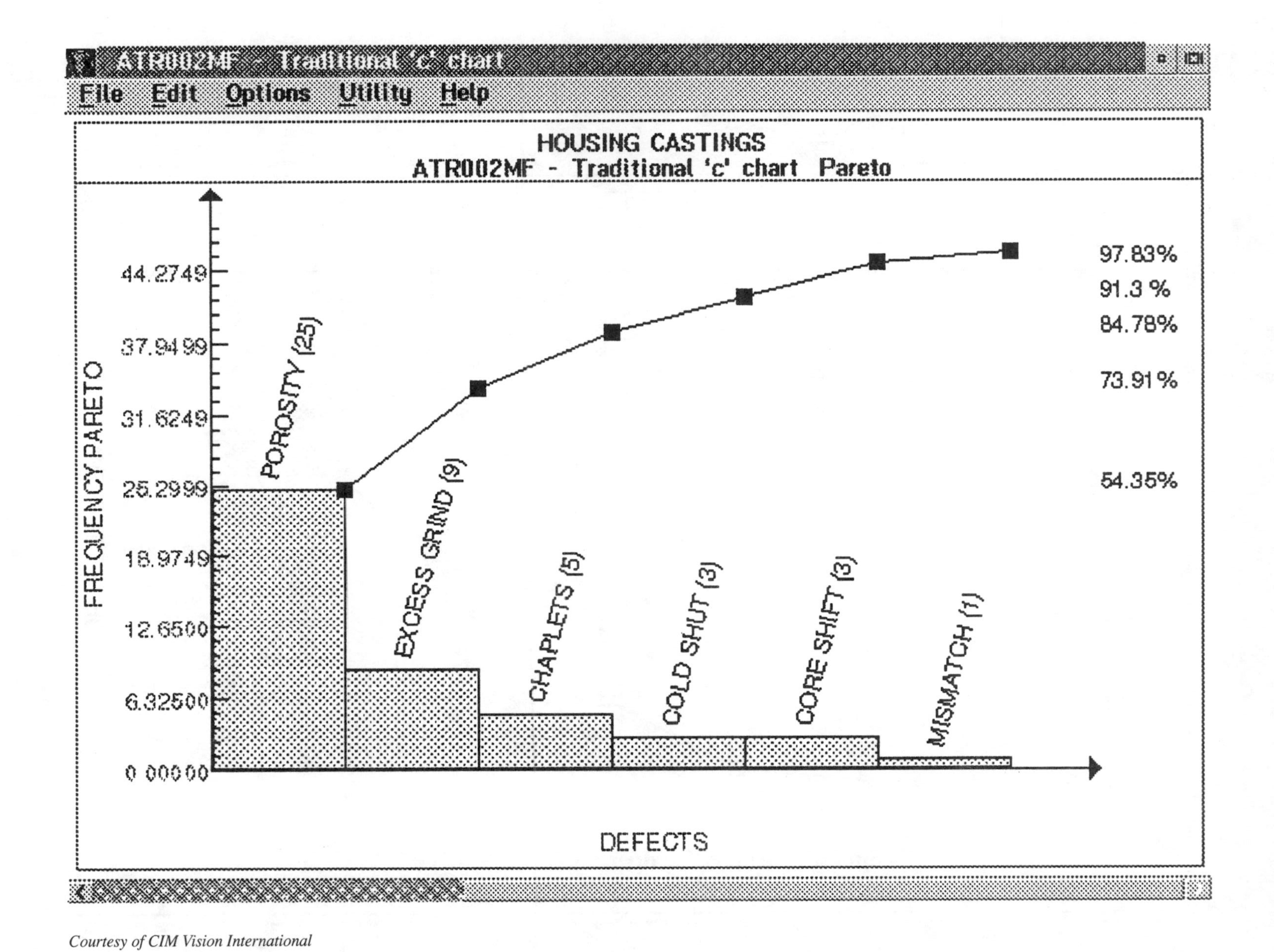

Courtesy of CIM Vision International

Figure 10.5. SQM software example of a frequency Pareto analysis.

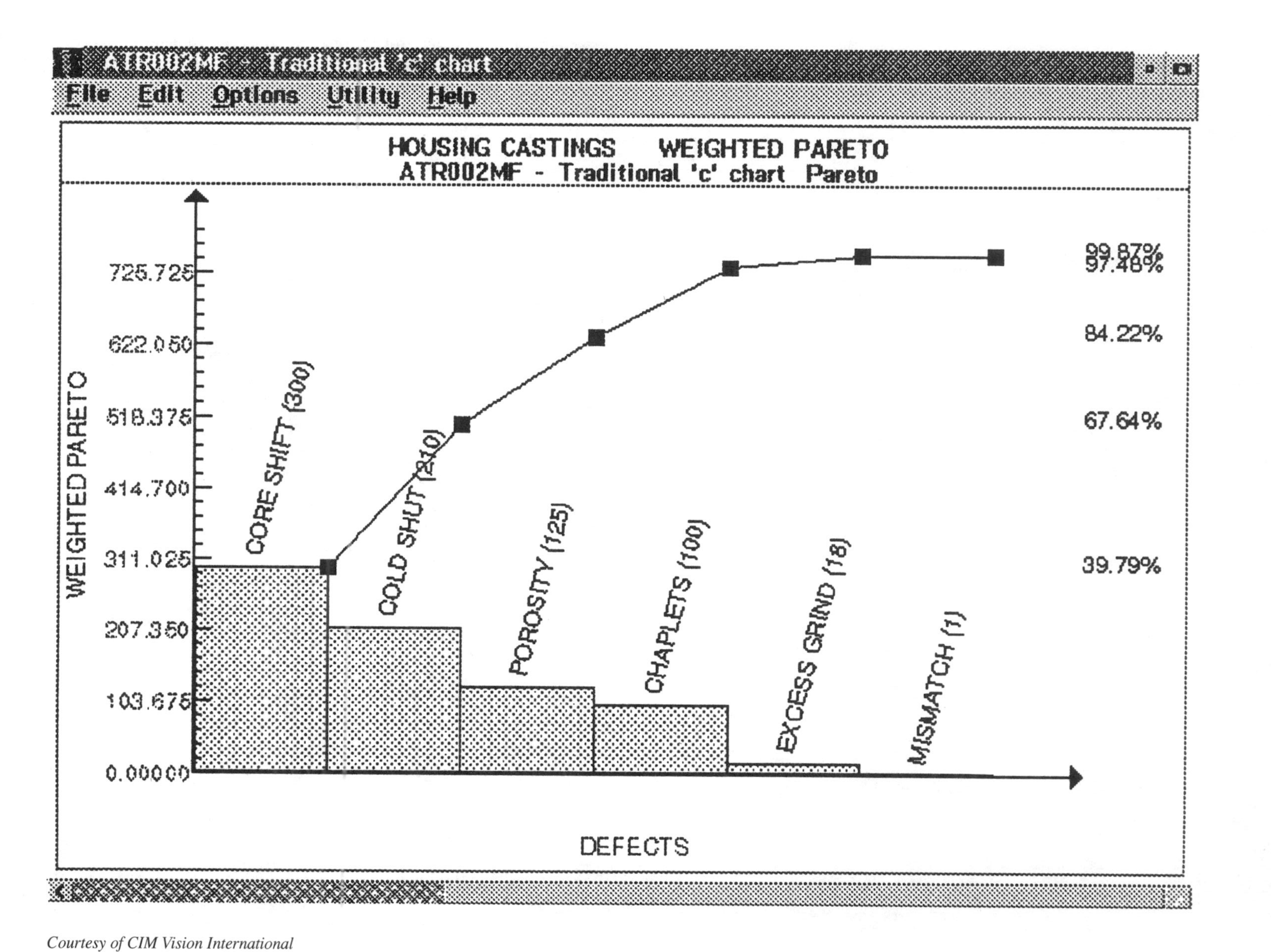

Courtesy of CIM Vision International

Figure 10.6. SQM software example of a weighted Pareto analysis.

Decision matrix			
Criteria \ Alternatives	A	B	C
Yield (C_{pk})	10	18	35
Production rate	4	12	18
Cycle time	8	8	12
Cost per part	2	4	22
Voting: 1 = poor 3 = average 5 = best			

Figure 10.7. Matrix diagram (alternative C is best).

Take Action. The responsible group should then take the action selected. Specific individual assignments should be given along with realistic dates for completion. Other departments or groups that may be affected by the action(s) should be advised or even participate so that the communication loop is closed.

Follow Up on the Action Taken

Reasons for Follow-Up. Follow-up is the key to effective problem solving for many reasons. Follow-up includes

1. Ensuring that the problem has been eliminated
2. Verifying that the true cause was corrected
3. Ensuring that the action taken was correct

Follow Up on the Specific Action. Make sure that the action was taken as planned and the expected results exist. This can be done by witnessing the action, performing a follow-up audit, or providing other objective forms of proof.

Follow Up on the Problem. Using the same data (reports, statistical analysis, Pareto analysis, and so on) a comparison can be made on the problem severity before and after the action was taken to see if the action caused favorable results. If so, it is evident that the root cause was corrected.

Monitor or Control the Level Achieved

Once an acceptable level has been achieved (to the standard), it is good practice to monitor or control that level in order to maintain it, improve upon it, and to detect potential new causes before they occur. Various methods can be used to monitor or control these levels. These, among others, include

- Control charts
- Histograms

- Pareto analysis
- Reports (rejection, scrap, rework, and so forth)
- Audits

To summarize, problem solving can be effective when effective methods are used to detect, analyze, act, and control the problem. The tools for problem solving and control can be statistical or not depending on the nature of the problem and the data necessary for solution.

OTHER PROBLEM-SOLVING TOOLS

Check Sheets

A check sheet (see Figure 10.8) can be used to list specific items such as defects, defectives, or other information for data collection purposes. Check sheets are quantitative in that they can be used to collect data in terms of counts. They provide a format for collecting predetermined data. Check sheets can be constructed in a manner that will automatically result in a histogram or Pareto chart.

Flowcharts

In some cases, the flow (or order of steps in a process) can be the root cause of the problem. Some examples are

- Operations in poor sequence
- Too many operations (or steps)
- Missing operations
- Redundant operations

In other cases, flowcharts help users understand the process so that problem causes can be isolated and eliminated. Flowcharts (see Figure 10.9) are helpful to help identify the ordered sequence of events that lead to a given output.

Coupler assemblies		**Assembly line defects**
Defect	**Count**	**Total**
Misaligned	~~////~~ ~~////~~	10
Wrong clamps	~~////~~ ~~////~~ ~~////~~ ~~////~~ ~~////~~	25
Loose bolts	~~////~~	5
Bad seal	~~////~~ ~~////~~ //	12
Missing lock wire	~~////~~ ~~////~~	10
Failed test	~~////~~	5
No nameplate	~~////~~ //	7
	Grand total	74

Figure 10.8. Check sheet for assembly line defects.

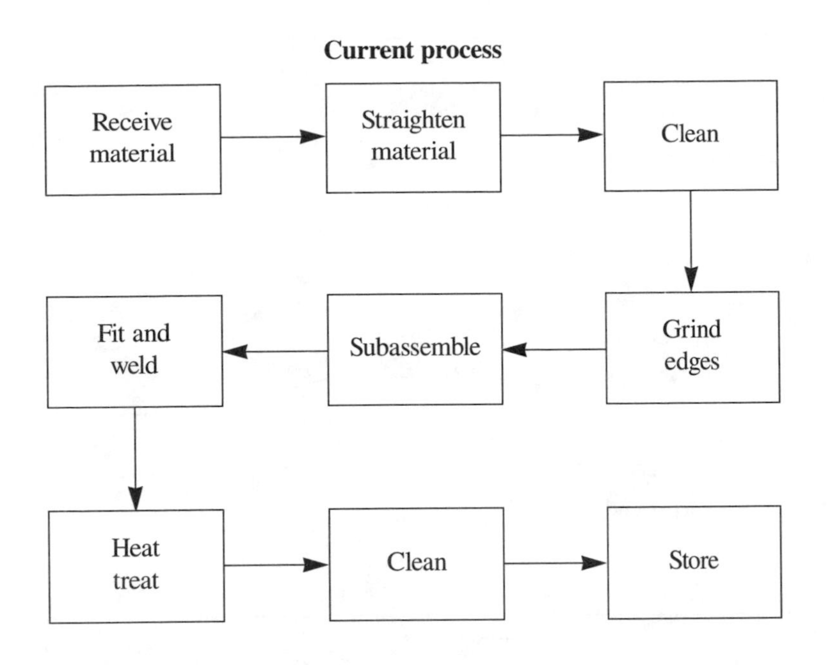

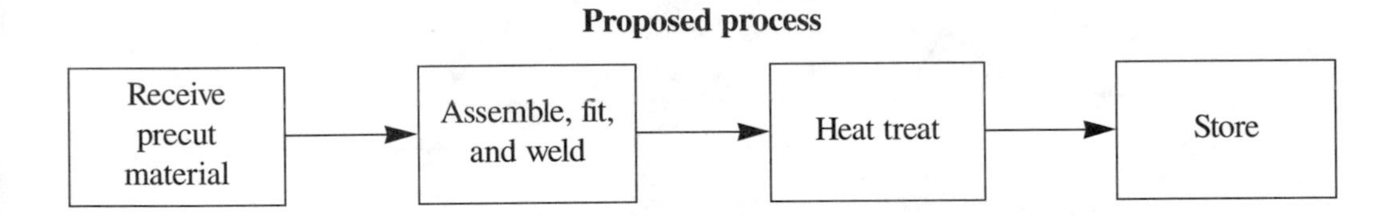

Figure 10.9. Flowchart comparing current process to proposed process.

Flowcharts help to identify

- Problems with the order of operations
- Nonvalue-added steps
- The relationship among steps, and their relative importance
- The difference between current processing steps and proposed steps

Run Charts

A *run chart* (see Figure 10.10) displays measurements or observations of a variable over a specified period of time (or order). Run charts can be used to identify process problems such as tool wear, die wear, and so on because upward or downward trends can be seen. Run charts are often used as an initial method for collecting data from a process until enough data can be collected to calculate control limits. In these cases, the run chart becomes a control chart.

Control Charts

A statistical *control chart* (Figure 10.11) is a tool that is used to identify special (assignable) causes of variation in a process at the time they occur. Correcting such special causes

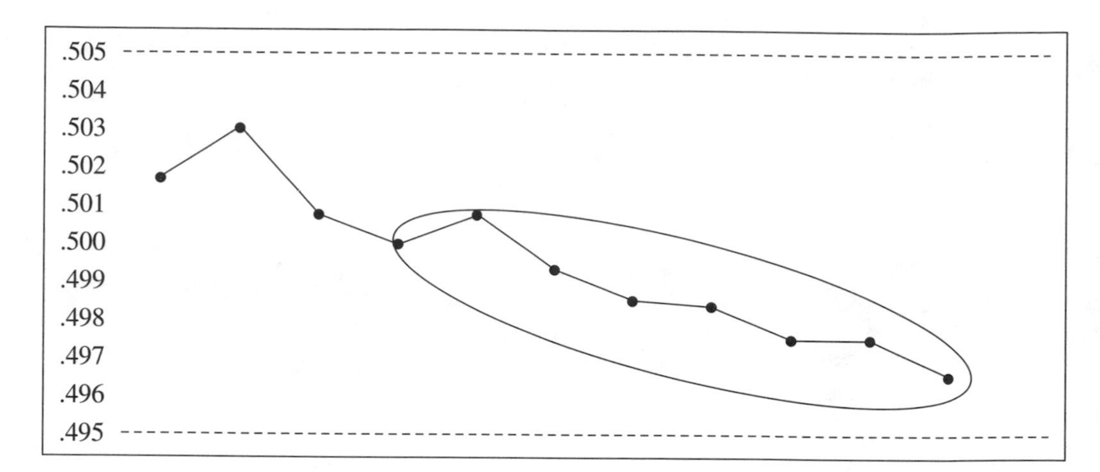

Figure 10.10. Run chart (trend).

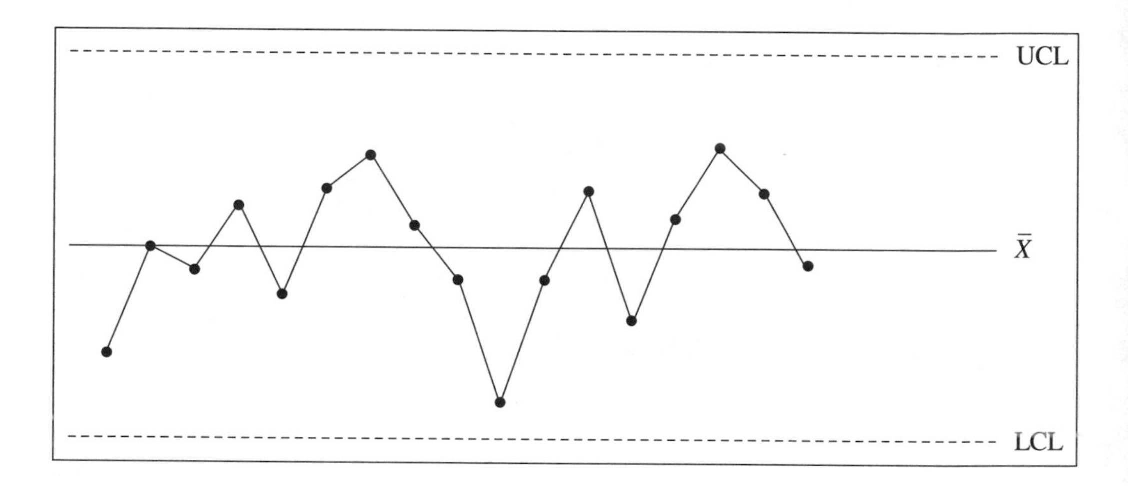

Figure 10.11. Control chart.

helps to bring the process in control. Control charts help users understand and control the variation in any process. Refer to chapters 2 through 6 for detailed information.

Histograms

Histograms (see Figure 10.12) are graphs that display the central tendency and variation of a process. The horizontal axis of a histogram indicates the scale for the individual measurements, and the vertical axis indicates the frequency of occurrence of each measurement. The histogram helps users to understand

- The process location (using the mean)
- The process variation
- The shape of the distribution of the process (for example, normal)

For further information on histograms, refer to chapter 7.

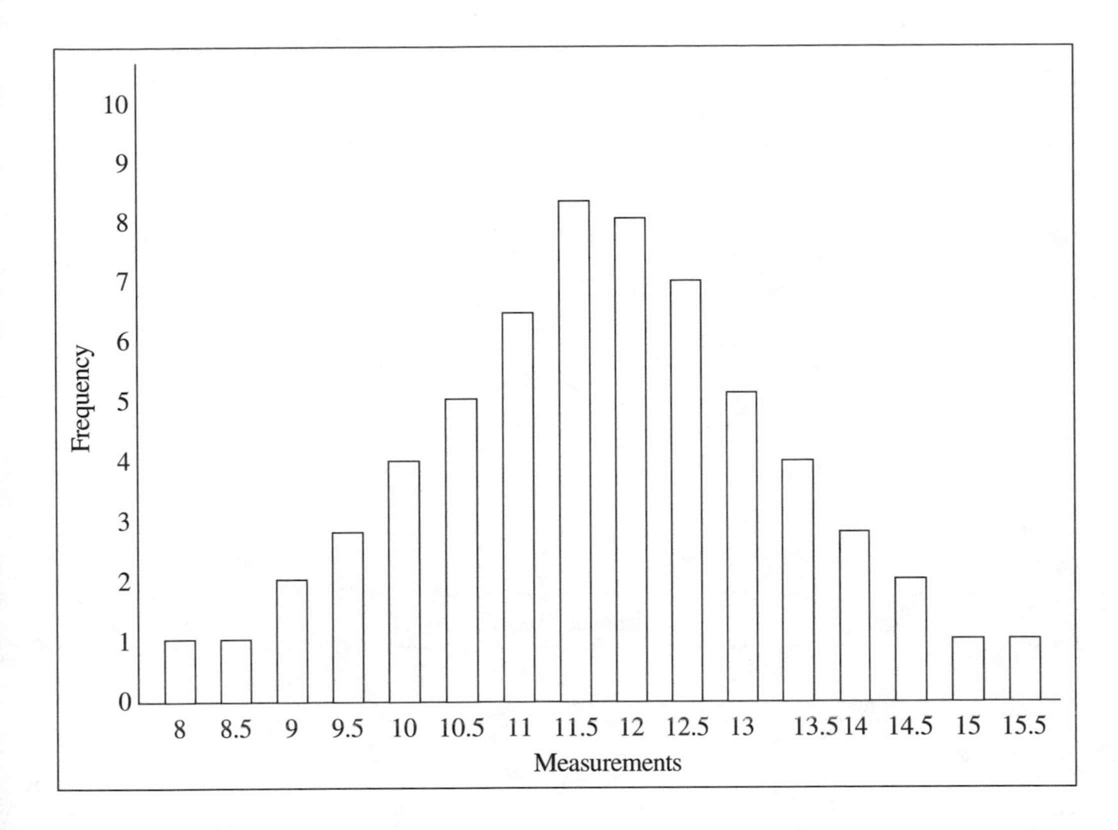

Figure 10.12. Histogram with normal distribution.

Scatter Diagrams

Scatter diagrams (see Figure 10.13) are used to graphically show the relationship between two variables (called *X* and *Y*). If two variables are assumed to be related, scatter diagrams help to see if there is a relationship and to test the strength of the relationship. The independent variable (*X*) is compared to the dependent variable (*Y*). Many problems (involving two variables) can be solved by using the knowledge that they are, or are not, related. For example,

- The *Y* variable cannot be measured, but the *X* variable can be measured. If they correlate, *X* measurements can be used to verify *Y* measurements.
- The *Y* variable must be measured by destroying the part where the *X* variable can be measured without destroying the part. If they correlate, nondestructive *X* measurements can be used to verify the *Y* variable.

For further information on scatter diagrams, regression, and correlation, see chapter 9.

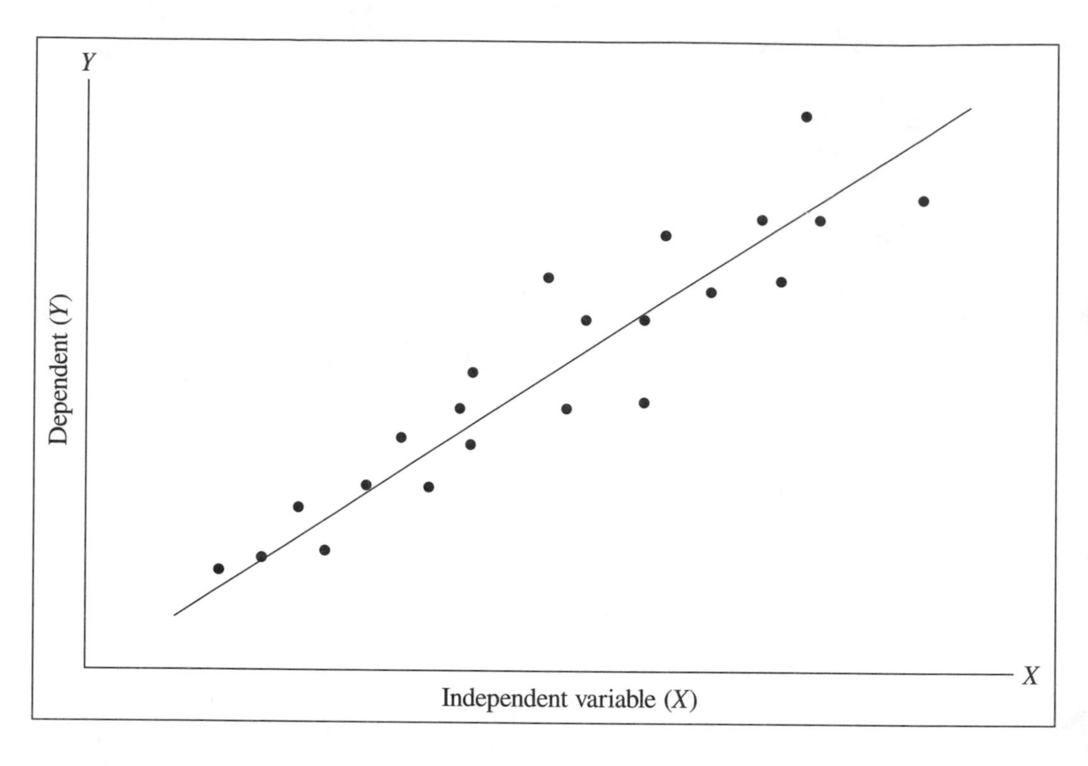

Figure 10.13. Scatter diagram showing positive correlation.

REVIEW PROBLEMS

Refer to Appendix H for solutions.

1. A problem can only be solved by eliminating the ______________ cause.
2. The six categories of causes of problems are manpower, methods, materials, machines, measurement, and ______________.
3. One problem-solving tool that can be used (along with brainstorming) to help structure the causes is the ______________ and ______________ diagram.
4. One of the keys to effective problem solving is to ______________ ______________ on the action taken to make sure that the action was effective in eliminating the root cause.
5. The group who should be formed to solve a problem should consist of those who are closest to the problem (or, in other words, the problem should select the people). (True or false)
6. A useful tool for gathering specific predetermined data at any stage of a process is called the ______________ .
7. ______________ charts help us to identify nonvalue-added operations, redundant operations, or out-of-sequence operations.
8. Several alternative actions and the effects of those actions can be identified by using a ______________ .

9. A useful tool for spotting trends in a process (by plotting individual data in order of production) is the ______________ chart.
10. A chart that is used to identify special (assignable) causes of variation in a process is called a ______________ chart.
11. A tool that shows the central tendency and variation of a process is the ______________.
12. A tool that is used to see if there is a relationship between two variables is the ______________ diagram.
13. A histogram shows the ______________ of occurrence of each value on the horizontal scale of measurements.
14. The tool that shows the shape of the distribution of individuals data is called a ______________.
15. Which of the following has upper and lower control limits?
    a. Run chart
    b. Matrix
    c. Scatter diagram
    d. Control chart

CHAPTER 11

Miscellaneous Topics

WHEN DEFECTS ARE FOUND IN A SUBGROUP

In cases where defects are found in the subgroup on a control chart, action should be taken as follows:

1. Identify and segregate the defects that have been found.
2. Screen all products made since the last time the control chart was in control.
3. Take action to correct the process.
4. Verify that the action taken corrected the problem, and continue the process.

ASSEMBLY CONTROL WITH SPC

Control charts can be used in assembly for assembly characteristics. One must realize that, in assembly, the purpose of the control chart is to control only the characteristics that are created or required at the assembly level. Often, a chart in assembly is used and the special cause of variation is with the detail parts. The detail parts are the building blocks of assembly dimensions. For example, Part A, B, and C are fastened together to create an assembly dimension. The overall length of the assembly requires control. How can it be controlled? Control charts should be placed during the process on each detail part dimension that is a building block to the assembly dimension.

Another example is an important characteristic created in assembly (such as torque on the fasteners) needs to be controlled. The solution is to apply control chart techniques to control that characteristic on the assembly line.

SPECIAL PROCESS CONTROL

Special processes such as plating, painting, heat treat, welding, brazing, and many others are often controlled effectively by monitoring the process parameters instead of the process output. For example, a heat treat process requires control. The control on this process could be concentrated on the important parameters of heat treating such as: temperature, time, and quench time.

In processes where the percent solution is a major factor on whether the output will be good or not, process control charts can ensure that solution tanks are properly controlled.

Many problems occur in processes due to overcontrol or undercontrol. Overcontrol occurs when solution is added to a tank that it doesn't need it. Undercontrol occurs when solution is not added to a tank that does need it.

Process control charts have applications in all special processes and can be an important factor toward getting the required results or successful output.

SEQUENTIAL CONTROLS

One must realize that every operation in the company has a customer. Material variation at any point in the process can cause major problems in subsequent operations. It is always wise to review the process flow and identify certain operations that produce characteristics that are important to the success of a future operation.

For example, operation 10 machines a diameter that is used as a locating diameter at operation 30. If this diameter has wide variation, it can cause operation 30 to vary also. The operator should use SPC on operation 10 to ensure the success of future operation, and therefore ensure that the end product meets specifications.

In another example, castings are used for the product and purchased from a vendor. There are characteristics in castings that have an important bearing on the success of machining operations such as: harness, machine stock, and target. The casting supplier can use SPC to control important casting characteristics and avoid casting variations that increase the variation of the machining process.

OPERATIONAL CAPABILITY ANALYSIS

Operational capability analysis is a technique that can be used to identify specific operations that are the source of problems. The method involves process capability studies on a serialized lot of parts at each operation involved in producing the final product or among a group of suspect operations. The analysis of involved operations could be performed by team brainstorming or flowcharting, or, in some cases, the operations are obvious. Consider the following case study.

Case Study

Final inspection (operation 80) was finding a high level of scrap and rework from a specific product line. The failure costs of quality were unacceptable. One Pareto analysis indicated that 80 percent of the scrap and rework across the entire product line was dimensional. Another Pareto analysis indicated specific dimensions A, C, E, and F were the top contributors. Cause codes for the scrap and rework were not consistent (did not help identify the source of the scrap and rework). These codes seemed to indicate that operations 40, 50, 60, and 70 were all suspect, and operation 40 was the operation that was taking most of the blame for the scrap and rework. The following steps outline the operational capability analysis used to solve this problem.

Step 1. Isolate the operations involved. A problem-solving team used a flowchart and their experience with the process. The team noted on the flowchart that the defective dimensions did not exist before machining operation 40 and were being found at final inspection operation 70. Therefore, specific operations 40, 50, 60, and 70 were suspect.

Step 2. Select a sample of serialized parts for study. A sample of 50 parts was selected out of a production run that had just been completed at operation 30 and was ready to go to operation 40. The team had all of the parts numbered 1 through 50 for the study. Numbering parts helps to compare parts across all operations, if necessary, and, if any measurements are questionable, find the part and measure it again.

Step 3. Perform capability studies after each operation. The parts were put through each operation (40 through 70) as they would normally be produced. There was no special treatment during processing. After each operation was completed, capability studies were performed on all four dimensions prior to further processing.

Step 4. Analyze the results. In this particular case, operation 40 was no longer suspect because capability studies indicated a centered, normally distributed, and capable process. Operation 50, of course, showed some change in the analysis of the four dimensions, but the change was not significant enough to cause the levels of scrap and rework. The yield of each dimension at operation 50 was still high, and the distributions were still normal. Operation 60, however, showed a drastic change in the central tendency, normality, and yield results. Operation 60 was the source of the problem. At this point, the team stopped the study and decided not to study operation 70 because the type of processing was not expected to change the dimensions in question.

Operational capability analysis, in this case, was very effective. The team further studied operation 60 and found that the operation was not very well planned and certain process settings were incorrect. The people at the operation were performing the operation as it was planned and running the process according to the specified settings. Once the settings were analyzed and improved, further capability analysis showed that the process was now in control and capable.

In this case, once operation 60 was made controllable and capable, the high levels of scrap and rework went down immediately and significantly. Note also that operational capability analysis proved that the prime suspect operation (40) was not the source of the problem.

VARIOUS CONTROL CHARTING ERRORS TO AVOID

Control charts can be misplotted for several different reasons. Plotting errors on a control chart will cause poor decisions to be made on the operation of the process. Some specific errors to avoid are

Measurement errors, including selecting the wrong instrument for the measurement, choosing the wrong discrimination, selecting an uncalibrated instrument, using an incorrect measurement method, not being a trained observer, and recording imprecise data.

Miscalculations, such as the charts are prepared manually and the person miscalculates the averages, medians, ranges, moving ranges, and so on. These errors can be avoided by using SPC software.

Misplotting, such as using one chart for two different processes, entering biased data, plotting the point in the wrong place, and so on. Some plotting errors can be avoided by using SPC software.

Data entry errors (using SPC software), such as poor setup of the chart file, typing errors (for example, decimal point in the wrong place), bias in the data being entered, and choosing a software product that will not perform the proper chart. SPC software can help solve many of the aforementioned errors, but some new ones exist.

SELECTING SUBGROUP SIZES AND INTERVALS

No absolute rule can be used for how often to sample. The realities of the factory layout, production run size, and the cost of sampling must be balanced with the value of the data obtained. In general, it is best to sample at close intervals at the beginning and increase the interval when the process results permit. Table 11.1 can be used for estimating the amount and frequency of sampling required. For example, a process produces 3000 parts per shift. Using Table 11.1, consider sampling 50 parts during the shift. If the control chart uses subgroups of 5 each, then $^{50}/_{5} = 10$; therefore 10 subgroups should be taken during the shift. On an eight-hour shift, samples would be taken every 48 minutes or so.

SAMPLE SIZE SELECTION PRECAUTIONS

Control limits are affected by the sample size. For example, smaller sample sizes cause wider control limits, and larger sample sizes can cause narrower control limits. One may choose to start out with larger sample sizes for increased ability to detect process shifts and later decrease the sample size as better control is achieved.

PROCESS CAPABILITY AND PERCENT YIELD INDICATORS

The most popular indicators of process capability and percent yield today are

C_p—Process capability index

$$C_p = \frac{\text{Total part tolerance}}{6\sigma}$$

Table 11.1. Samples selected based on production rate.

Production rate/shift	Total number of parts to sample per shift
1 to 65	5
66 to 110	10
111 to 180	15
181 to 300	25
301 to 500	30
501 to 800	35
801 to 1300	40
1301 to 3200	50
3201 to 8000	60
8001 to 22,000	85

C_{pk}—Worst case capability index

C_{pk} (for bilateral specifications)

$$C_{pk} = \text{Lesser of } = \frac{USL - \bar{X}}{3\sigma} \text{ or } \frac{\bar{X} - LSL}{3\sigma}$$

C_{pk} (for maximum limit specifications)

$$C_{pk} = \frac{USL - \bar{X}}{3\sigma}$$

C_{pk} (for minimum limit specifications)

$$C_{pk} = \frac{\bar{X} - LSL}{3\sigma}$$

Percent yield is a measure of the percent of good products that are expected to be made the first time by the process. Percent yield applies to variables or attributes data. The percent yield is computed using Z scores and the table of areas under the normal curve (Appendix A).

PPM (defective parts per million) is a measure of the estimated number of defectives expected from a process if one million parts were made.

$$PPM = \text{Percent defective} \times 10{,}000$$

These measures of capability have already been covered in the appropriate chapters, but there is a direct relationship between them that should be understood. Examples of this relationship are shown in the Table 11.2. The percent yield and PPM values in Table 11.2 are all equivalents of the worst case (C_{pk}) and the associated percent yield.

The C_p index is defined as the specification tolerance width divided by the process six-sigma spread irrespective of process centering. This is the best that can possibly be obtained with the current process. The C_{pk} is the capability index that accounts for process centering error. It relates to the distance between the process mean and the nearest specification limit divided by one-half of the process spread (3σ).

Table 11.2. Examples of equivalence for capability indicators.

C_p	C_{pk}	Sigma capability	Percent yield*	PPM*	Remarks
1.0	0.5	1.5 sigma	93.32%	66,800	Process is not centered, and not capable.
1.0	1.0	3 sigma	99.73%	2700	Process is centered, but not capable.
1.33	1.33	4 sigma	99.994%	63	Process is centered and capable.
2.0	1.33	4 sigma	99.994%	63	Process is not centered, but is capable.
2.0	2.0	6 sigma	99.9994%	.002	Process is well centered and very capable.

*Based on the Table of Areas Under the Normal Curve (Appendix A)

When processes are centered precisely on the mid-value of specifications (for example, nominal), the C_p and C_{pk} will be the same value. If the process is off center, the C_{pk} index will be lower than the C_p index.

The yield of a process is determined using the table of areas under the normal curve (Appendix A). For any given C_{pk}, there is an equivalent percent yield. The parts per million defective, then, is computed based on 100 percent minus the percent yield times 10,000.

RATIONAL SUBGROUPS FOR CONTROL CHARTS

A very important aspect of control charting is to collect data from the process in rational subgroups in accordance with the purpose of the chart. A *subgroup* is a sample of consecutively produced products that are made during a given time period and taken from the process at specific time intervals. A *rational subgroup* is subgroup samples containing consecutively made products that are taken from the process that one would expect would have little variation, yet there is noticeable variation from subgroup to subgroup. Due to the purpose of most statistical control charts (previously mentioned), the most rational subgroups are taken in order of production. But there is much more to rational subgrouping than just the order of production. The purpose of most statistical control charts is to detect assignable (special) causes of variation in the process and to eliminate those causes during production. Those special causes can be associated with three types of variation, and rational subgrouping depends on the types of variation one intends to control.

Lot-to-lot variation or the variability between lots or batches. Examples are different lots with different material hardness or different tooling is used on each different lot that affects the setup.

Part-to-part variation is the variation between parts (or subgroups of parts). Examples are the natural variation of a process or external noise variables that affect the next part or the next subgroup of parts.

Within-the-part variation is the variation within one part. Examples are a tapered shaft and an out-of-round bore.

The objective of process control is, of course, to be able to consistently identify when special causes of variation are present in a process, and to take specific action to eliminate (or remove) those special causes when they occur. Rational subgrouping will help provide the operator with correct signals of when the process goes out of control, and training and support will help the operator to correct those causes when they occur and ensure that the control chart will help achieve the objective. This section is intended to point out that rational subgrouping is an important aspect of any process control plan that involves statistical control charts. For more detailed information on rational subgrouping, refer to the reference texts in the bibliography.

Example of Within-the-Piece Variation

A process is producing multiple similar dimensions, and the variation between the dimensions, on one part (within-the-piece variation), is expected to be significant. An example of this is a press that punches three different holes (A, B, and C) in each part, in three different places, with three different punches. The holes all have the same nominal dimension and specification limits. If it was expected that the different punches would wear at different rates (which could happen), the three different holes measured per part would have significant within-the-piece variation and they would *not* be a rational subgroup to control the press. It may be necessary to construct three different charts. If the three different holes, A, B, and C, were measured and were treated as one subgroup on one chart, the variation within the piece would not only cause false signals on that control chart, but the operator would be less likely to see the wear of any of the individual punches.

A second example is: a dimension on a sheet metal part is being formed, and the dimension varies within the part from end to end. It would not be a rational subgroup if operators measure the same part in three places and call those three measurements a subgroup. Rational subgrouping would cause the measurements to be taken at one place on each part, or a separate chart is prepared for each end.

A lathe is turning shafts where the outside diameter of the shaft is the selected control characteristic. If the lathe is in good condition where the shafts were not expected to have significant taper from end to end in the diameter, then subgroup measurements could be made anywhere on the shaft and still be rational subgroups. However, if the lathe is expected (or known) to produce significantly tapered shafts a rational subgroup would be one where all operators measure the parts at the same location on the shaft or different charts are prepared (one for the shaft diameter at one designated end and one for the shaft diameter at the opposite end). If the shafts were significantly tapered, and the measurements were at random places, this within-the-piece variability would cause false signals on the control chart, hence false actions (or the lack of action).

Example 3

Another example (similar to example 2) is a boring process that is known (or expected) to produce inside diameters that are significantly out-of-round. If they are significantly out-of-round, it is best to simply measure the variation in one specified location of the inside diameter. Many may be tempted to measure the bore in four places, then compute the average diameter of each bore, and combine the averages into a subgroup. This irrational type of subgrouping can hide the variation (make it look a lot smaller than it is) and cause poor process control results.

Example of Lot-to-Lot Variation

Various different material lots are produced on a process in which each lot has several sheets of metal. The variation within the sheet and from sheet to sheet is consistent, but the variation from lot to lot is not. A rational subgroup would be one that signaled the lot-to-lot variation (if that were the purpose of the chart). If subgroup samples are taken only at random, this would not be a rational subgroup and may cause the process to appear to be in control when it is not.

Time and the Order of Production

Rational subgroups should also be subgroups that are representative of production during a given period of time, therefore it is important that the time period between subgroups is identified in the process control plan, then followed by all operators running the process. In this manner, when the process goes out of control (special cause(s) are present), the operator can take action to identify and eliminate the cause(s). There are two typical errors that are made, at times, by operating personnel during control charting. These errors will always cause trouble in the process control effort. They are

1. Operators take the data at the right time intervals, but actually plot the data at the end of the shift. I call this the too-late syndrome. By the time the operator sees the pattern of plot points and any out-of-control patterns, it is too late to do anything about them. They have already had their adverse affect on the products made during the shift.
2. Operators do not take the data at the correct time intervals. Data are taken on the last few parts at the end of the shift. I call this the wasted-your-time syndrome.

Errors Caused by Irrational Subgroups

There are a host of errors (or problems) caused by taking irrational subgroups from a process. Two of the most significant errors are the Type I (producer's risk) and Type II (consumer's risk) errors on control charts. Basically, these errors are

Type I (producer's risk). The risk associated with false signals on the control chart that make the process appear to be out of control when it is not. Hence, action is taken on the process that should not have been taken.

Type II (consumer's risk). The risk associated with false signals on the control chart that make the process appear to be in control when it is really out of control. Hence, no action is taken to correct the special cause(s) and control/reduce the variation of the process.

During the original process control planning for SPC on any given process, the process should be investigated to the extent that rational subgroups are taken in line with the purpose of the control chart.

REVISING STATISTICAL CONTROL LIMITS

All of the statistical control charts covered in this book (except for the standardized charts) require a review of the control limits for possible revision. This is so because these charts are *standard-not-given charts,* which means that the control limits are computed using data from the process. There are many different viewpoints of SPC practitioners about when control limits should be revised. Some of the differences have to do with the type of chart being used, the objective of the chart, and the chosen level of sensitivity for the process control system. For these same reasons there are some who do not use some of the tests for out-of-control patterns covered in chapter 6. If the goal, however, is continuous variability reduction, then the control limits of the process should always represent the actual common cause system of variation in the process. In this case, revisions to chart centerlines and control limits will be made for reasons to follow.

The following are examples of when control limits should be reviewed for possible revision and the reasons for revising the limits.

1. When the process goes out of control and assignable (special) causes have been corrected, the centerlines and control limits should be revised using the last 20 subgroups that are in control (excluding the plot points and raw data that are associated with the out-of-control pattern). Control limits should not be revised if the cause has not been corrected.

2. When there has been a significant improvement in variability, the centerlines and control limits should be revised for the process because a different common cause system of variability has been achieved.
3. When the variability of the process has worsened significantly, the centerlines and control limits should be revised for the process because a different common cause system of variability has resulted.
4. When historical data have been used to start the chart, then sufficient data have been retrieved to revise the centerlines and control limits. This, of course, applies when the user starts the chart with historical data intending to update the limits when enough recent data have been collected. It also applies when hypothesis tests confirm that the recent data collected from the process is not representative of the historical data that were used from the start.

If centerlines and control limits are not revised at the right time for the right reasons, they will not properly represent the common cause system of variability of the process and can be the cause of Type I and Type II errors on control charts. Keep in mind that process variability is not static, it's dynamic. The variability of a process changes (for better or worse) as the process itself changes.

NOTES TAKEN ON CONTROL CHARTS

Control charts are a running record of the variability of a process over time that help the processor to understand more about process variability, normal and abnormal. Notes should be taken on the chart (or some other record) so that the effect of special causes, adjustments, changes in methods, and many other variables can be understood. The more effective operators become in taking notes on the process, the more chance they will have in controlling and reducing unwanted variability.

The process engineer can also use notes taken on the process in order to correlate process actions with the resulting behavior of variability from that action. For example, a machine shop operator may be interested in cutting tool life and the effect of cutting tool life on variability. First of all, any machine shop is interested in increasing the yield of product from every cutting tool, but does not want to risk extending the use of a worn tool that may cause defective products. Tool wear is often seen as some degree of trend on the control chart. The trend may be gradual or steep, but tool wear is a typical cause for trends.

If one were to monitor the gradual trend and the histogram for central tendency and spread, one could tell how long the use of the tool could be extended before reaching the probability of defective products. This can and has been achieved by simply taking notes and watching the control chart and histogram for the results. Actual cases in which this has been done used statistical software so that it was easy and fast to switch from viewing the control chart to the histogram and back.

Notes should be taken, at minimum, in the following cases.

1. When adjustments are made to the process, it should be noted on the control chart so that the effect of the adjustment can be seen. At times, certain adjustments cause the process to go out of control. At other times, the adjustment was necessary and correct, and this is confirmed by the fact that process regained control status after the adjustment was made.

2. Whenever the process goes out of control, notes should be taken that indicate what action was taken to bring the process back into control. This helps correlate the action that was taken to the status of control to see if it was successful. It also provides feedback to the customer so that there is proof that a serious process control effort is being made.
3. When there is a change in the method used to manufacture the product, charts should be noted because the method change must be confirmed to be the cause of improvement or worsening variability.
4. Whenever there is defective product found in a subgroup (even if the process is in statistical control), notes should be taken regarding the cause of the defective(s) and the action taken to correct the problem and/or sort the product.

REVIEW PROBLEMS

Refer to Appendix H for solutions.

1. The best approach toward controlling an assembly stack-up dimension is to
   a. Measure and chart the dimension in the assembly area
   b. Tighten the tolerances of the individual parts
   c. Control the processes involved with each stack-up dimension
   d. None of the above
2. In cases where special processes (such as plating, heat treating, and so on) are involved, it is often most effective to control specific process ________________ that have a direct affect on the desired output variables.
3. If the success of a future operation depends on a dimension that is produced at an earlier operation, it follows that one should control that dimension at the earlier operation. (True or false)
4. Which of the following types of errors can be expected when plotting a control chart manually?
   a. Measurement errors
   b. Miscalculations
   c. Misplotting
   d. All of the above
5. Which of the following types of errors can be expected when using software to prepare and maintain control charts?
   a. Miscalculations
   b. Misplotting
   c. Measurement
   d. Data entry
   e. c and d

6. Operational capability analysis is an experiment that involves the production of a serialized lot of product through different operations for the purpose of isolating the source of unwanted variability. (True or false)

7. A process control chart is being used to monitor the outside diameter of a shaft. The specifications for the shaft are .500–.510 diameter. The process average is .504." The standard deviation of the process is .001." Answer the following questions regarding the capability of the process.

   a. What is the C_p index for the process?

   b. What is the C_{pk} index for the process?

   c. Calculate the percent yield for the process.

   d. Compute the equivalent PPM for the process.

APPENDIX A

Table of Areas Under the Normal Curve

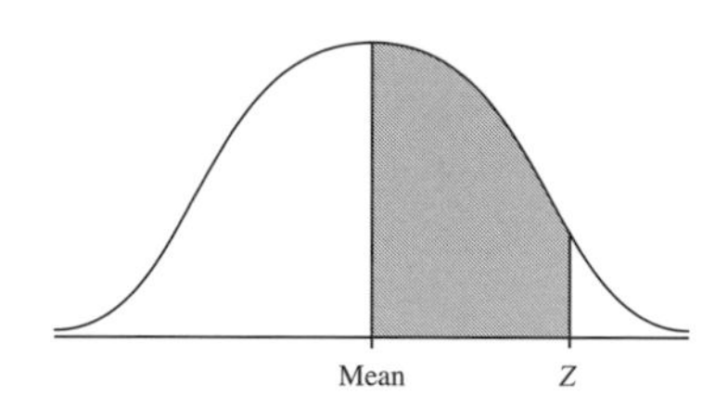

Z	0	1	2	3	4	5	6	7	8	9
0.0	.00001	.0040	.0080	.0120	.0160	.0199	.0239	.0279	.0319	.0359
0.1	.0398	.0438	.0478	.0517	.0557	.0596	.0636	.0675	.0714	.0753
0.2	.0793	.0832	.0871	.0910	.0948	.0987	.1026	.1064	.1103	.1141
0.3	.1179	.1217	.1255	.1293	.1331	.1368	.1406	.1443	.1480	.1517
0.4	.1554	.1591	.1628	.1664	.1700	.1736	.1772	.1808	.1844	.1879
0.5	.1915	.1950	.1985	.2019	.2054	.2088	.2123	.2157	.2190	.2224
0.6	.2257	.2291	.2324	.2357	.2389	.2422	.2454	.2486	.2518	.2549
0.7	.2580	.2612	.2642	.2673	.2704	.2734	.2764	.2794	.2823	.2852
0.8	.2881	.2910	.2939	.2967	.2995	.3023	.3051	.3078	.3106	.3133
0.9	.3159	.3186	.3212	.3238	.3264	.3289	.3315	.3340	.3365	.3389
1.0	.3413	.3438	.3461	.3485	.3508	.3531	.3554	.3577	.3599	.3621
1.1	.3643	.3665	.3686	.3708	.3729	.3749	.3770	.3790	.3810	.3830
1.2	.3849	.3869	.3888	.3907	.3925	.3944	.3962	.3980	.3997	.4015
1.3	.4032	.4049	.4066	.4082	.4099	.4115	.4131	.4147	.4162	.4177
1.4	.4192	.4207	.4222	.4236	.4251	.4265	.4279	.4292	.4306	.4319
1.5	.4332	.4345	.4357	.4370	.4382	.4394	.4406	.4418	.4429	.4441
1.6	.4452	.4463	.4474	.4484	.4495	.4505	.4515	.4525	.4535	.4545
1.7	.4554	.4564	.4573	.4582	.4591	.4599	.4608	.4616	.4625	.4633
1.8	.4641	.4649	.4656	.4664	.4671	.4678	.4686	.4693	.4699	.4706
1.9	.4713	.4719	.4726	.4732	.4738	.4744	.4750	.4756	.4761	.4767
2.0	.4772	.4778	.4783	.4788	.4793	.4798	.4803	.4808	.4812	.4817
2.1	.4821	.4826	.4830	.4834	.4838	.4842	.4846	.4850	.4854	.4857
2.2	.4861	.4864	.4868	.4871	.4875	.4878	.4881	.4884	.4887	.4890
2.3	.4893	.4896	.4898	.4901	.4904	.4906	.4909	.4911	.4913	.4916
2.4	.4918	.4920	.4922	.4925	.4927	.4929	.4931	.4932	.4934	.4936
2.5	.4938	.4940	.4941	.4943	.4945	.4946	.4948	.4949	.4951	.4952
2.6	.4953	.4955	.4956	.4957	.4959	.4960	.4961	.4962	.4963	.4964
2.7	.4965	.4966	.4967	.4968	.4969	.4970	.4971	.4972	.4973	.4974
2.8	.4974	.4975	.4976	.4977	.4977	.4978	.4979	.4979	.4980	.4981
2.9	.4981	.4982	.4982	.4983	.4984	.4984	.4985	.4985	.4986	.4986
3.0	.4986500	.4990323	.4993128	.4995165	.4996630	.4997673	.4998409	.4998922	.4999276	.4999519
4.0	.4999683	.4999793	.4999866	.4999915	.4999940	.4999968	.4999979	.4999987	.4999992	.4999995
5.0	.4999997	.4999998	.4999999	.4999999	.500000	.500000	.500000	.500000	.500000	.500000

Example: A *Z* value of 2.35 equals .4906 (or 49.06 percent of the area under the curve).

APPENDIX B

Factors for Control Charts

Sample size n	A_2	A_3	A_5	$\bar{\tilde{A}}_2$	d_2	D_3	D_4	D_5	D_6	c_4	B_3	B_4	E_2
2	1.880	2.659	2.224	1.88	1.128	0	3.267	0	3.865	.798	0	3.267	2.66
3	1.023	1.954	1.265	1.19	1.693	0	2.574	0	2.745	.886	0	2.568	1.77
4	0.729	1.628	.829	0.80	2.059	0	2.282	0	2.375	.921	0	2.266	1.46
5	0.577	1.427	.712	0.69	2.326	0	2.114	0	2.179	.940	0	2.089	1.29
6	0.483	1.287	.562	0.55	2.534	0	2.004	0	2.055	.952	.030	1.970	1.18
7	0.419	1.182	.520	0.51	2.704	.076	1.924	.078	1.967	.959	.118	1.882	1.11
8	0.373	1.099	.441	0.43	2.847	.136	1.864	.139	1.901	.965	.185	1.815	1.05
9	0.337	1.032	.419	0.41	2.970	.184	1.816	.187	1.850	.969	.239	1.761	1.01
10	0.308	0.975	.369	0.36	3.078	.223	1.777	.227	1.809	.973	.284	1.716	0.98

APPENDIX C

Table of Bias Factors (Attribute Gage Studies)

B_{Fa} and B_{Miss} are based on computed P_{Fa} and P_{Miss} values, respectively.

P_{Fa} P_{Miss}	B_{Fa} B_{Miss}	P_{Fa} P_{Miss}	B_{Fa} B_{Miss}
8.01	.0264	0.26	.3251
0.02	.0488	0.27	.3312
0.03	.0681	0.28	.3372
0.04	.0863	0.29	.3429
0.05	.1040	0.30	.3485
0.06	.1200	0.31	.3538
0.07	.1334	0.32	.3572
0.08	.1497	0.33	.3621
0.09	.1626	0.34	.3668
0.10	.1758	0.35	.3712
0.11	.1872	0.36	.3739
0.12	.1989	0.37	.3778
0.13	.2107	0.38	.3814
0.14	.2227	0.39	.3836
0.15	.2323	0.40	.3867
0.16	.2444	0.41	.3885
0.17	.2541	0.42	.3910
0.18	.2613	0.43	.3925
0.19	.2709	0.44	.3945
0.20	.2803	0.45	.3961
0.21	.2874	0.46	.3970
0.22	.2966	0.47	.3977
0.23	.3034	0.48	.3984
0.24	.3101	0.49	.3989
0.25	.3187	0.50	.3989

APPENDIX D

Fisher *F* Test for Variances

Short-run target charts require that the variability among several different parts numbers charted together is representative. In chapter 3, two tests for representative variability were given: the range test and the *F* test. The *F* test is one of the options that can be used to test the variability between two samples. This test is more mathematical than the range test, but it is also more effective in isolating nonrepresentative parts. The *F* ratio is used to compare the variances between two sets of sample data to see if they are representative or not.

The *F* test is based on an educated guess, or hypothesis. In this case there are two possible hypotheses: H_0 (the null hypothesis) or H_1 (the alternate hypothesis). In *F* testing, the null hypothesis is that the variability of the samples is representative. The alternate hypothesis is that the variability of the samples is not representative.

Steps to Construct a Fisher F *Test*

The following steps for performing the *F* test will also serve as an example problem for practical purposes.

1. Calculate the required statistics for both samples. The following statistics are needed for each sample to perform the *F* test.

   s = sample standard deviation

   V = variance (which is equal to s^2)

   n = the sample size

   Sample 1 $s_1 = 12.57$ $V_1 = 158.0$ $n_1 = 30$

   Sample 2 $s_2 = 11.51$ $V_2 = 132.5$ $n_2 = 30$

2. Establish the hypothesis. The hypothesis here is H_0, the two samples are representative.

3. Calculate the *F* ratio.

$$F_{\text{ratio}} = \frac{\text{Variance}_{\text{largest}}}{\text{Variance}_{\text{smallest}}}$$

   Note: If both variances are equal, the *F* ratio is 1. The *F* ratio can never be less than 1 because the numerator is always the largest variance.

For example,

$$F_{ratio} = \frac{158.0}{132.5} = 1.19$$

4. Establish the degrees of freedom (*df*) and alpha risk. Refer to the *F* table in Appendix E.
   - The numerator (largest variance) and denominator (smallest variance) are each $n = 30$, so the degrees of freedom $(n - 1)$ is 29.
   - The alpha risk for this example is set at 5 percent (or .05) for 95 percent confidence. The alpha risk could be selected for higher confidence.
5. Find the cutoff value (refer to the *F* table in Appendix E) and perform the hypothesis test. Using the degrees of freedom applied to the numerator and denominator (which is the sample size minus 1), obtain the cutoff value from the table (1.85 in this case). See Figure D.1.
   - If the actual *F* ratio (1.19 in this case) is higher than the cutoff value (1.85), there is significant difference in the two variances and the alternate hypothesis (H_1) is true.
   - If the actual *F* ratio is lower than the cutoff value, there is no significant difference in the two variances.

In this example, the actual *F* ratio (1.19) is less than the cutoff value (1.85), therefore there is no significant difference in the variances of the two samples. If these two samples were charted on a short-run target chart, this indicates that the part is representative and charting it on this chart can continue.

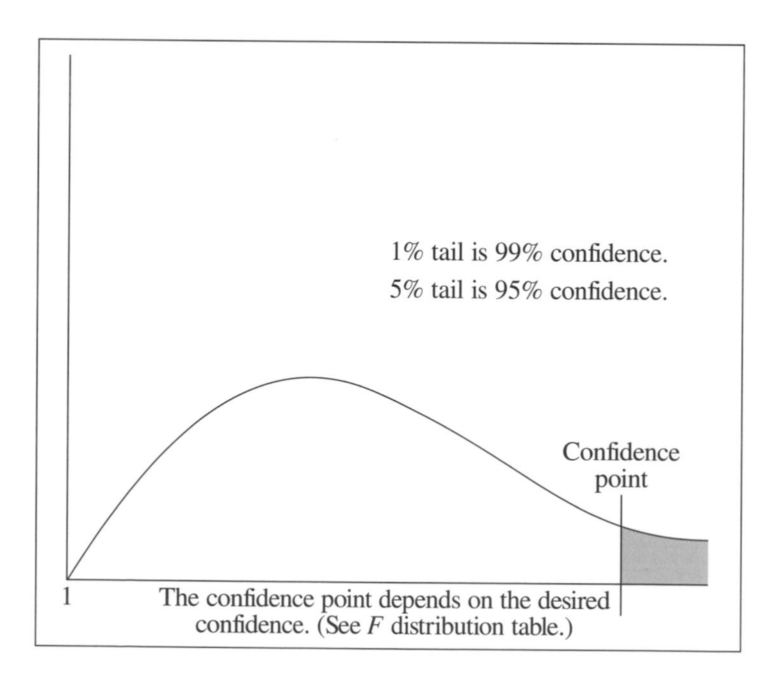

Figure D.1. The confidence point for *F* test.

APPENDIX E

F Distribution

Degrees of freedom (*df*) of the greater variance (numerator).

	1		2		3		4		5		6	
$p \rightarrow$	.05	.01	.05	.01	.05	.01	.05	.01	.05	.01	.05	.01
1	161	4,052	200	4,999	216	5,403	225	5,625	230	5,764	234	5,859
2	18.51	98.49	19.00	99.01	19.16	99.17	19.25	99.25	19.30	99.30	19.33	99.33
3	10.13	34.12	9.55	30.81	9.28	29.46	9.12	28.71	9.01	28.24	8.94	27.91
4	7.71	21.20	6.94	18.00	6.59	16.69	6.39	15.98	6.26	15.52	6.16	15.21
5	6.61	16.26	5.79	13.27	5.41	12.06	5.19	11.39	5.05	10.97	4.95	10.67
6	5.99	13.74	5.14	10.92	4.76	9.78	4.53	9.15	4.39	8.75	4.28	8.47
7	5.59	12.25	4.74	9.55	4.35	8.45	4.12	7.85	3.97	7.46	3.87	7.19
8	5.32	11.26	4.46	8.65	4.07	7.59	3.84	7.01	3.69	6.63	3.58	6.37
9	5.12	10.56	4.26	8.02	3.86	6.99	3.63	6.42	3.48	6.06	3.37	5.80
10	4.96	10.04	4.10	7.56	3.71	6.55	3.48	5.99	3.33	5.64	3.22	5.39
11	4.84	9.65	3,98	7.20	3.59	6.22	3.36	5.67	3.20	5.32	3.09	5.07
12	4.75	9.33	3.88	6.93	3.49	5.95	3.26	5.41	3.11	5.06	3.00	4.82
13	4.67	9.07	3.80	6.70	3.41	5.74	3.18	5.20	3.02	4.86	2.92	4.62
14	4.60	8.86	3.74	6.51	3.34	5.56	3.11	5.03	2.96	4.69	2.85	4.46
15	4.54	8.68	3.68	6.36	3.29	5.42	3.06	4.89	2.90	4.58	2.79	4.32
16	4.49	8.53	3.63	6.23	3.24	5.29	3.01	4.77	2.85	4.44	2.74	4.20
17	4.45	8.40	3.59	6.11	3.20	5.18	2.96	4.67	2.81	4.34	2.70	4.10
18	4.41	8.28	3.55	6.01	3.16	5.09	2.93	4.58	2.77	4.25	2.66	4.01
19	4.38	8.18	3.52	5.93	3.13	5.01	2.90	4.50	2.74	4.17	2.63	3.94
20	4.35	8.10	3.49	5.85	3.10	4.94	2.87	4.43	2.71	4.10	2.60	3.87
21	4.32	8.02	3.47	5.78	3.07	4.87	2.84	4.37	2.68	4.04	2.57	3.81
22	4.30	7.94	3.44	5.72	3.05	4.82	2.82	4.31	2.66	3.99	2.55	3.76
23	4.28	7.83	3.42	5.66	3.03	4.76	2.80	4.26	2.64	3.94	2.53	3.71
24	4.26	7.82	3.40	5.61	3.01	4.72	2.78	4.22	2.62	3.90	2.51	3.67
25	4.24	7.77	3.38	5.57	2.99	4.68	2.76	4.18	2.60	3.86	2.49	3.63
26	4.22	7.72	3.37	5.53	2.98	4.64	2.74	4.14	2.59	3.82	2.47	3.59
27	4.21	7.68	3.35	5.49	2.96	4.60	2.73	4.11	2.57	3.79	2.46	3.56
28	4.20	7.64	3.34	5.45	2.95	4.57	2.71	4.07	2.56	3.76	2.44	3.53
29	4.18	7.60	3.33	5.42	2.93	4.54	2.70	4.04	2.54	3.73	2.43	3.50
30	4.17	7.56	3.32	5.39	2.92	4.51	2.69	4.02	2.53	3.70	2.42	3.47
32	4.15	7.50	3.30	5.34	2.90	4.46	2.67	3.97	2.51	3.66	2.40	3.42
34	4.13	7.44	3.28	5.29	2.88	4.42	2.65	3.93	2.49	3.61	2.38	3.38
38	4.10	7.35	3.25	5.21	2.85	4.34	2.62	3.86	2.46	3.54	2.35	3.32
42	4.07	7.27	3.22	5.15	2.83	4.29	2.59	3.80	2.44	3.49	2.32	3.26
46	4.05	7.21	3.20	5.10	2.81	4.24	2.57	3.76	2.42	3.44	2.30	3.22
50	4.03	7.17	3.18	5.06	2.79	4.20	2.56	3.72	2.40	3.41	2.29	3.18
60	4.00	7.08	3.15	4.98	2.76	4.13	2.52	3.65	2.37	3.34	2.25	3.12
80	3.96	6.96	3.11	4.88	2.72	4.04	2.48	3.56	2.33	3.25	2.21	3.04
100	3.94	6.90	3.09	4.82	2.70	3.98	2.46	3.51	2.30	3.20	2.19	2.99
200	3.89	6.76	3.04	4.71	2.65	3.88	2.41	3.41	2.26	3.11	2.14	2.90
1000	3.85	6.68	3.00	4.62	2.61	3.80	2.38	3.34	2.22	3.04	2.10	2.82
∞	3.84	6.64	2.99	4.60	2.60	3.78	2.37	3.32	2.21	3.02	2.09	2.80

(*Continued*)

	7		8		10		12		16		20	
$p \rightarrow$	.05	.01	.05	.01	.05	.01	.05	.01	.05	.01	.05	.01
1	237	5928	239	5981	242	6056	244	6106	246	6169	248	6208
2	19.36	99.34	19.37	99.36	19.39	99.40	19.41	99.42	19.43	99.44	19.44	99.45
3	8.88	27.67	8.84	27.49	8.78	27.23	8.74	27.05	8.69	26.83	8.66	26.69
4	6.09	14.98	6.04	14.80	5.96	14.54	5.91	14.37	4.85	14.15	5.80	14.02
5	4.88	10.45	4.82	10.27	4.74	10.05	4.68	9.89	4.60	9.68	4.56	9.55
6	4.21	8.26	4.15	8.10	4.06	7.87	4.00	7.72	3.92	7.52	3.87	7.39
7	3.79	7.00	3.73	6.84	3.63	6.62	3.57	6.47	3.49	6.27	3.44	6.15
8	3.50	6.19	3.44	6.03	3.34	5.82	3.28	5.67	3.20	5.48	3.15	5.36
9	3.29	5.62	3.23	5.47	3.13	5.26	3.07	5.11	2.98	4.92	2.93	4.80
10	3.14	5.21	3.07	5.06	2.97	4.85	2.91	4.71	2.82	4.52	2.77	4.41
11	3.01	4.88	2.95	4.74	2.86	4.54	2.79	4.40	2.70	4.21	2.65	4.10
12	2.92	4.65	2.85	4.50	2.77	4.30	2.69	4.16	2.60	3.98	2.54	3.86
13	2.84	4.44	2.77	4.30	2.67	4.10	2.60	3.96	2.51	3.78	2.46	3.67
14	2.77	4.28	2.70	4.14	2.50	3.94	2.53	3.80	2.44	3.62	2.39	3.51
15	2.70	4.14	2.64	4.00	2.55	3.80	2.48	3.67	2.39	3.48	2.33	3.36
16	2.68	4.03	2.59	3.89	2.49	3.69	2.42	3.55	2.33	3.37	2.28	3.25
17	2.63	3.93	2.55	3.79	2.45	3.59	2.38	3.45	2.20	3.27	2.23	3.16
18	2.58	3.85	2.51	3.71	2.41	3.51	2.34	3.37	2.25	3.19	2.19	3.07
19	2.55	3.77	2.48	3.63	2.38	3.43	2.31	3.30	2.21	3.12	2.15	3.00
20	2.52	3.71	2.45	3.56	2.35	3.37	2.28	3.23	2.18	3.05	2.12	2.94
21	2.49	3.65	2.42	3.51	2.32	3.31	2.25	3.17	2.15	2.99	2.09	2.88
22	2.47	3.59	2.40	3.45	2.30	3.26	2.23	3.12	2.13	2.94	2.07	2.83
23	2.45	3.54	2.38	3.41	2.28	3.21	2.20	3.07	2.10	2.89	2.04	2.78
24	2.43	3.50	2.36	3.36	2.26	3.17	2.18	3.03	2.09	2.85	2.02	2.74
25	2.41	3.46	2.34	3.32	2.24	3.13	2.16	2.99	2.06	2.81	2.00	2.70
26	2.39	3.42	2.32	3.29	2.22	3.09	2.15	2.96	2.05	2.77	1.99	2.66
27	2.37	3.39	2.30	3.26	2.20	3.06	2.13	2.93	2.03	2.74	1.97	2.63
28	2.36	3.36	2.29	3.23	2.19	3.03	2.12	2.90	2.02	2.71	1.96	2.60
29	2.35	3.33	2.28	3.20	2.18	3.00	2.10	2.87	2.00	2.68	1.94	2.57
30	2.34	3.30	2.27	3.17	2.16	2.98	2.09	2.84	1.99	2.66	1.93	2.55
32	2.32	3.25	2.25	3.12	2.14	2.94	2.07	2.80	1.97	2.62	1.91	2.51
34	2.30	3.21	2.23	3.08	2.12	2.89	2.05	2.76	1.95	2.58	1.89	2.47
38	2.26	3.15	2.19	3.02	2.09	2.82	2.02	2.69	1.92	2.51	1.85	2.40
42	2.24	3.10	2.17	2.96	2.06	2.77	1.99	2.64	1.89	2.46	1.82	2.35
46	2.22	3.05	2.14	2.92	2.04	2.73	1.97	2.60	1.87	2.42	1.80	2.30
50	2.20	3.02	2.13	2.88	2.02	2.70	1.95	2.56	1.85	2.39	1.78	2.26
60	2.17	2.93	2.10	2.82	1.99	2.63	1.92	2.50	1.81	2.32	1.75	2.20
80	2.12	2.87	2.05	2.74	1.95	2.55	1.88	2.41	1.77	2.24	1.70	2.11
100	2.10	2.82	2.03	2.69	1.92	2.51	1.35	2.36	1.75	2.19	1.68	2.06
200	2.05	2.73	1.98	2.60	1.87	2.41	1.80	2.28	1.69	2.09	1.62	1.97
1000	2.02	2.66	1.95	2.53	1.84	2.34	1.76	2.20	1.65	2.01	1.58	1.89
∞	2.01	2.64	1.94	2.51	1.83	2.32	1.75	2.18	1.64	1.99	1.57	1.87

(Continued)

→	30		50		100		∞	
p	.05	.01	.05	.01	.05	.01	.05	.01
1	230	6258	252	6302	253	6334	254	6368
2	19.46	99.47	19.47	99.48	19.49	99.49	19.50	99.50
3	8.62	26.50	8.58	26.35	8.56	26.23	8.53	26.12
4	5.74	13.83	5.70	13.69	5.66	13.57	5.63	13.46
5	4.50	9.38	4.44	9.24	4.40	9.13	4.36	9.02
6	3.81	7.23	3.75	7.09	3.71	6.99	3.67	6.88
7	3.38	5.98	3.32	5.85	3.28	5.75	3.23	5.63
8	3.08	5.20	3.03	5.06	2.98	4.96	2.93	4.86
9	2.86	4.64	2.80	4.51	2.76	4.41	2.71	4.31
10	2.70	4.25	2.64	4.12	2.59	4.01	2.54	3.91
11	2.57	3.94	2.50	3.80	2.45	3.70	2.40	3.60
12	2.46	3.70	2.40	3.56	2.35	3.46	2.30	3.36
13	2.38	3.51	2.32	3.37	2.26	3.27	2.21	3.16
14	2.31	3.34	2.24	3.21	2.19	3.11	2.13	3.00
15	2.25	3.20	2.18	3.07	2.12	2.97	2.07	2.87
16	2.20	3.10	2.13	2.96	2.07	2.86	2.01	2.75
17	2.15	3.00	2.08	2.86	2.02	2.76	1.96	2.65
18	2.11	2.91	2.04	2.78	1.98	2.68	1.92	2.57
19	2.07	2.84	2.00	2.70	1.94	2.60	1.88	2.49
20	2.04	2.78	1.96	2.63	1.90	2.53	1.84	2.42
21	2.00	2.72	1.93	2.58	1.87	2.47	1.81	2.36
22	1.98	2.67	1.91	2.53	1.84	2.42	1.78	2.31
23	1.96	2.62	1.88	2.48	1.82	2.37	1.76	2.26
24	1.94	2.58	1.86	2.44	1.80	2.33	1.73	2.21
25	1.92	2.54	1.84	2.40	1.77	2.29	1.71	2.17
26	1.90	2.50	1.82	2.36	1.76	2.23	1.69	2.13
27	1.88	2.47	1.80	2.33	1.74	2.21	1.67	2.10
28	1.87	2.44	1.78	2.30	1.72	2.18	1.65	2.06
29	1.85	2.41	1.77	2.27	1.71	2.15	1.64	2.03
30	1.84	2.38	1.76	2.24	1.69	2.13	1.62	2.01
32	1.82	2.34	1.74	2.20	1.67	2.08	1.59	1.96
34	1.80	2.30	1.71	2.15	1.64	2.04	1.57	1.91
38	1.76	2.22	1.67	2.08	1.60	1.97	1.53	1.84
42	1.73	2.17	1.64	2.02	1.57	1.91	1.49	1.78
46	1.71	2.13	1.62	1.98	1.54	1.86	1.46	1.72
50	1.69	2.10	1.60	1.94	1.52	1.82	1.44	1.68
60	1.65	2.03	1.56	1.87	1.48	1.74	1.39	1.60
80	1.60	1.94	1.51	1.78	1.42	1.65	1.32	1.49
100	1.57	1.89	1.48	1.73	1.39	1.59	1.28	1.43
200	1.52	1.79	1.42	1.62	1.32	1.48	1.19	1.28
1000	1.47	1.71	1.36	1.54	1.26	1.38	1.08	1.11
∞	1.46	1.69	1.35	1.52	1.24	1.36	1.00	1.00

APPENDIX F

Sample Process Control Plan

DoRight Mfg. Co.

Process:	Part name:	Part no.:	
Next assembly:	Prepared by:	Approved by:	Date prepared:
Team leader:	Title:	Title:	Date revised:

Control characteristics

A. Outside diameter	D.	G.	
B. Runout to datum -A-	E.	H.	
C.	F.	I.	

Process control plan

Control characteristics					Measurements		Process variation		
Char.	Specification limits	Statistical method	Sample size and frequency	Initial C_{pk}	Gage/ method	Gage R&R	Process step/ operator number	Key process parameter and setting	Control method

APPENDIX G

Glossary for Statistical Quality Control

Arithmetic mean and arithmetic average (μ, $\bar{X}$, or $\bar{Y}$): (1) $\bar{X}$ or $\bar{Y}$ (X-bar or Y-bar) Sample average (2) μ population mean. A measure of central tendency or location that is the sum of the observations divided by the number of observations. *Note:* There are several averages such as the arithmetic average, geometric average, and so on.

Assignable cause: A factor that contributes to variation and is feasible to detect and identify.

Attributes, method of: Measurement of quality by the method of attributes consists of noting the presence (or absence) of some characteristic or attribute in each of the units in the group under consideration and counting how many units do (or do not) possess the quality attribute or how many such events occur in the unit, group, or area. *Note:* One of the most common attribute measures for acceptance sampling is the percentage of nonconforming units.

c (Count): The count or number of events of a given classification occurring in a sample. More than one event may occur in a unit (area of opportunity), and each such event throughout the sample is counted.

Central line: A line on a control chart representing the long-term average or a standard value of the statistical measure being plotted.

Chance causes (random causes): Factors, generally numerous and individually of relatively small importance, that contribute to variation, but are not feasible to detect or identify.

Characteristic: A property that helps to differentiate between items of a given sample or population. *Note:* The differentiation may be either quantitative (by variables) or qualitative (by attributes).

Coefficient of variation: A measure of relative dispersion that is the standard deviation divided by the mean and multiplied by 100 to give a percentage value. This measure cannot be used when the data take both negative and positive values or when they have been coded in such a way that the value $X = 0$ does not coincide with the origin.

Excerpted from *Glossary and Tables for Statistical Quality Control* with permission of ASQC.

Control chart: A graphical method for evaluating whether a process is or is not in a state of statistical control. (The determinations are made through comparison of the values of some statistical measure(s) for an ordered series of samples, or subgroups, with control limits.) *Note:* A control chart is used to make decisions about a process. There are a variety of specific charts, each designed for the type of decisions to be made, the nature of the data, and the type of statistical measure used. The term *control chart* was for many years synonymous with *Shewhart control chart* being based on the original work by W. A. Shewhart. There are now several other distinct types or major variations in use.

The emphasis of the control chart, as contrasted to the acceptance control chart, is on the process being in control rather than a direct evaluation of whether the measure of the product or service satisfies tolerance requirements. *Control chart,* however, is sometimes used in an acceptance sense calling for action or investigation when a process is deemed to have shifted from its standard level.

Control chart factor: A factor, usually varying with sample size, to convert specified statistics or parameters into a central line value or control limit appropriate to the control chart.

Control chart method: The method of using control charts to determine whether or not processes are in a stable state.

Control chart—standard given: A control chart whose control limits are based on adopted standard values applicable to the statistical measure plotted on the chart. *Note:* This type of control chart is used to discover whether observed values of $\bar{X}$, c_o, p_o, and so on, for samples differ from standard values $\bar{X}$, c_o, p_o, and so on, by an amount greater than should be attributed to chance. The standard value may be based on representative prior data or an economic value based on consideration of needs of service and cost of production or a desired or aimed-at value designated by a specification.

Control chart—no standard given: A control chart whose control limits are based on the sample or subgroup data plotted on the chart. *Note:* This type of control chart is used to determine whether observed values of $\bar{X}$, R, p, and so on, for a series of samples vary among themselves by an amount greater than should be attributed to chance. Control charts based entirely on the data from the samples being evaluated are used for detecting lack of constancy of the cause system. This type of chart is particularly useful in research and development stages to determine whether a new process or product is reproducible and whether test methods are repeatable.

Control limits: Limits on a control chart that are used as criteria for signaling the need for action or for judging whether a set of data does or does not indicate a state of statistical control. *Note:* When warning limits are used, the control limits are often called *action limits*. Action may be in the form of investigation of the source(s) of an assignable cause, making a process adjustment, or terminating a process. Criteria other than control limits are also used frequently.

Deviation (measurement sense): The difference between a measurement or quasi-measurement and its stated value or intended level. *Note:* A *deviation,* as used in a measurement sense, should be stated as a difference in terms of the appropriate data units. Sometimes these units will be original measurement units; sometimes they will be quasi-measurements (that is, a scaled rating of subjective judgments); sometimes they will be designated values representing all continuous or discrete measurements falling in defined cells or classes.

Frequency distribution: A set of all various values that individual observations may have and the frequency of their occurrence in the sample or population.

In-control process: A process in which the statistical measure(s) being evaluated are in a state of statistical control. *Note:* The term *process* may represent (a) the manufacture of physical and tangible products, (b) the output of services, (c) the collection of measurements, and (d) other variations such as paperwork.

Inspection: The process of measuring, examining, testing, gaging, or otherwise comparing the unit with the applicable requirements. *Note:* The term *requirements* sometimes is used broadly to include standards of good workmanship.

Kurtosis (γ_2 for populations, g_2 for samples): A measure of the shape of a distribution. A positive value of γ_2 indicates that the distribution has longer tails than the normal distribution (platykurtosis); while a negative value of γ_2 indicates that the distribution has shorter tails (leptokurtosis). For the normal distribution $\gamma_2 = 0$.

Lower control limit (LCL): Control limit for points plotting below the center level.

Median (Med): The middle measurement when an odd number of units are arranged in order of size; for an ordered set $X_1, X_2, \ldots, X_{2k-1}$.

$$Med = X_k$$

When an even number are so arranged, the median is the average of the two middle units; for an ordered set $X_1, X_2, \ldots, X_{2k}$.

$$Med = \frac{X_k + X_{k+1}}{2}$$

Mode: The most frequent value of the variable.

Nonconformity: A departure of a quality characteristic from its intended level or state that occurs with a severity sufficient to cause an associated product or service not to meet a specification requirement. *Note:* In some situations, specification requirements coincide with customer usage requirements (see definition of *defect*). In other situations they may not coincide, being either more or less stringent, or the exact relationship between the two is not fully known or understood. When a quality characteristic of a product or service is evaluated in terms of conformance to specification requirements, the use of the term *nonconformity* is appropriate. The emphasis on the work evaluated is that of making a decision concerning conformance, whereas an imperfection rating basically deals with a measurement process. Contractual obligations, stated or implicit, may be involved, or the specification requirements may be purely internal and deliberately set tighter than the customer requirements.

Nonconforming unit: A unit of product or service containing at least one nonconformity. *Note:* See *Nonconformity.*

np (number of affected units): The total number of units (areas of opportunity) in a sample in which an event of a given classification occurs. A unit (area of opportunity) is to be counted only once even if several events of the same classification are encountered therein.

p (1) (Used in the sense of a proportion or fraction) The ratio of the number of units (areas of opportunity) in which at least one event of a given classification occurs to the total number of units (areas of opportunity) sampled. A unit (area of opportunity) is to be counted only once even if several events of the same classification are encountered therein. (2) (Used in the sense of percent) The percentage of the total number of units (areas of opportunity) in a sample in which an event of a given classification occurs. A unit (area of opportunity) is to be counted only once even if several events of the same classification are encountered therein. *Note:* When *p* or the other statistical measures are used in analyses, it is customary to identify them by titles pertaining to the specific event.

Population: The totality of items or units of material under consideration. *Note:* The items may be units or measurements, and the population may be real or conceptual. Thus *population* may refer to all the items actually produced in a given day or all that might be produced if the process were to continue in-control.

Process capability: The limits within which a tool or process operates based upon minimum variability as governed by the prevailing circumstances. *Note:* The phrase "by the prevailing circumstances" indicates that the definition of inherent variability of a process involving only one operator, one source of raw material, and so on, differs from one involving multiple operators, many sources of raw material, and so on. If the measure of inherent variability (see *statistical control*) is made within very restricted circumstances, it is necessary to add components for frequently occurring assignable sources of variation that cannot economically be eliminated.

Quality: The totality of features and characteristics of a product or service that bear on its ability to satisfy given needs. *Note:* In order to be able to ensure, control, or improve quality, it is necessary to be able to evaluate it. This definition calls for the identification of those characteristics and features bearing upon the fitness-for-use of a product or service. The "ability to satisfy given needs" reflects value to the customer and includes economics as well as safety, availability, maintainability, reliability, design, and all other characteristics.

R-*Range:* A measure of dispersion that is the difference between the largest observed value and the smallest observed value in a given sample. While the range is a measure of dispersion in its own right, it is sometimes used to estimate the population standard deviation, but is a biased estimator unless multiplied by the factor ($1/d_2$) appropriate to the sample size. *Note:* Because the range tends to be inefficient, it is not recommended that ranges be used for large sample sizes. A rule of thumb suggests a maximum sample size of 10.

Formula: R = largest observation minus smallest observation

$$\text{Population estimate: } \hat{\sigma} = \frac{\overline{R}}{d_2}$$

Run: An uninterrupted sequence of occurrences of the same attribute or event in a series of observations, or a consecutive set of successively increasing (run up) or successively decreasing (run down) values in a series of variable measurements. *Note:* In control chart applications, some variable measurements are treated as attributes in determining runs. For example, a run might be considered a series of a specified number of points consequently plotting above or below the center level, or five consecutive points three of which fall outside of warning limits.

Sample: A group of units, portion of material, or observations taken from a larger collection of units, quantity of material, or observations that serves to provide information that may be used as a basis for making a decision concerning the larger quantity. *Note:* The sample may be the units or material themselves or the observations collected from them. The decision may or may not involve taking action on the units or material, or on the process.

Skewness (γ_1 for populations, g_1 for samples): A measure of the symmetry of a distribution. A positive value of γ_1 indicates that the distribution has a greater tendency to tail to the right (positively skewed or skewed to the right), and a negative value of γ_1 indicates a greater tendency of the distribution to tail to the left (negatively skewed or skewed to the left). For the normal distribution $\gamma_1 = 0$.

Specification limits: See *tolerance limits.*

Standard deviation: (1) s-population standard deviation is a measure of variability (dispersion) of observations that is the positive square root of the population variance. (2) s-sample standard deviation is a measure of variability (dispersion) of observations in the sample that is the positive square root of the sample variance.

$$s = \left[\frac{1}{n-1}\Sigma(X_i - \bar{X})^2\right]^{1/2} \text{ or } \left[\frac{1}{n-1}\Sigma(Y_i - \bar{Y})^2\right]^{1/2}$$

Note: From a simple random sample, s gives a biased estimate (underestimate) of σ. See factor C_4.

Standard error of the mean:

$$s_{\bar{X}} = \frac{Sx}{\sqrt{n}} \text{ or } s_{\bar{Y}} = \frac{Sy}{\sqrt{n}}$$

Standard error of c:

$$s_c = \sqrt{c}$$

Standard error of p:

$$s_p \cong S\sqrt{\frac{p\,(1-p)}{n}}$$

Standard error of the range:

$$S_R = \frac{(D_4 - 1)\,d_2\bar{R}}{3}$$

Standard error of the sample standard deviation:

$$s_s = s\sqrt{1 - c_4^2}$$

Standard error of the sample variance:

$$S_s^2 \cong S^2\left(\sqrt{\frac{2}{n-1}}\right)$$

Standard error of u *or* c/n:

where n is the number of items per unit

$$s_{c/kn} = \frac{1}{kn}\sqrt{c}$$

where k is the number of units in a group of units.

State of statistical control: A process is considered to be in a state of statistical control if the variations among the observed sampling results from it can be attributed to a constant system of chance causes.

Statistic: A quantity calculated from a sample of observations, most often to form an estimate of some population parameter. *Note:* Some statistics, such as the average ($\bar{X}$) and the sample variance (s^2), are unbiased estimators, in this instance of the mean (μ) and the variance (σ^2) respectively. Others, like the range (R) and sample standard deviation (s), are biased estimators, in this instance of the standard deviation (σ), and require a correction factor if the bias is to be removed. Bias often is of no practical concern particularly if it is constant and all comparisons are made on the same basis.

Subgroup: (1) (Object sense) A set of units or quantity of material obtained by subdividing a larger group of units or quantity of material. (2) (Measurement sense) A set of groups of observations obtained by subdividing a larger group of observations.

Tolerance (specification sense): The total allowable variation around a level or state (upper limit minus lower limit), or the maximum acceptable excursion of a characteristic *Note:* The determination of the amount of variation to be allowed involves the product or service requirements and consideration of process capability, measurement variability, and other appropriate elements or some compromise among these. See *tolerance limits*.

Tolerance limits (specification limits): Limits that define the nonconformance boundaries for an individual unit of a manufacturing or service operation. *Note:* Limits may be established either with or without the use of probability considerations. Tolerance limits may be in the form of a single (unilateral) unit (upper or lower) or double (bilateral) limits (upper and lower).

u *or* c/n *(count per unit):* The average count or average number of events of a given classification per unit (unit of opportunity) occurring within a sample. More than one event may occur in a unit (unit area of opportunity), and each such event is counted. *Note:* Sometimes $\bar{c}$ is used instead of c/n but it is necessary then to recognize that it is the $\bar{c}$ for the unit of n items and not for a group of k units as would be the case for c (count).

Upper control limit (UCL): Control limit for points plotting above the center level.

Variables, method of: Measurement of quality by the method of variables consists of measuring and recording the numerical magnitude of a quality characteristic for each of the units in the group under consideration. This involves reference to a continuous scale of some kind.

Variance: (1) σ^2—Population variance is a measure of variability (dispersion) of observations based upon the mean of the squared deviations from the arithmetic mean. *Note:* The population variance $\sigma^2 = \int_R (x - \mu)^2 f(x)dx$ where R is the region over which the random variable x is defined, where $f(x)$ is the probability density function, and where μ is the mean of $f(x)$. The population standard deviation (σ) is the square root of the population variance.

(2) s^2—Sample variance is a measure of variability (dispersion) of observations in a sample based upon the squared deviations from the arithmetic average divided by the degrees of freedom. *Note:* The sample variance s^2 should always be accompanied by its degrees of freedom. In this example, the degrees of freedom equal $(n - 1)$. From a simple random sample, s^2 gives an unbiased estimate of the population variance σ^2.

$$s^2 = \frac{1}{n - 1} \Sigma(X_i - \bar{X})^2 \text{ or } \frac{1}{n - 1} \Sigma(Y_i - \bar{Y})^2$$

(3) $s^3{}_{(ms)}$—Sample mean square division is a measure of variability of observations in a sample based upon the mean of the square deviations from the arithmetic average. *Note:* From a simple random sample, $s^2{}_{(ms)}$ gives a biased estimate (underestimate) of the population variance σ^2.

APPENDIX H

Answers to Chapter Review Problems

CHAPTER 2: TRADITIONAL VARIABLES CONTROL CHARTS

1.

		X		
	X	X	X	
	X	X	X	
60	65	70	75	80

2. Yes
3. .4996
4. .006
5. 59.8
6. 7
7. 51.15
8. 18.24
9. 0
10. 4.3
11. True
12. True
13. True
14. d
15. d
16. c
17. c
18. b
19. 3
20. .0022
21. .5018
22. .0026
23. .0006
24. .501
25. .752
26. 5 percent
27. .5001

CHAPTER 3: SHORT-RUN VARIABLES CONTROL CHARTS

1. 2.33
2. 12
3. 4.89
4. –.003
5. +.002
6. No
7. Process
8. 1
9. $+A_2$
10. +2.5

CHAPTER 4: TRADITIONAL ATTRIBUTES CONTROL CHARTS

1. 0.15
2. 0.097
3. 0.26
4. 15.2
5. 18
6. 0
7. 2.18
8. 0.24
9. 2.24
10. 16.65
11. c
12. c
13. c
14. b
15. d
16. Sporadic
17. Pareto
18. Sporadic
19. True
20. b
21. (a) 97.5 percent (b) 25,000 PPM

CHAPTER 5: SHORT-RUN ATTRIBUTES CONTROL CHARTS

1. Standardized
2. 0
3. UCL = +3 LCL = –3
4. +0.91
5. –1.5
6. +0.83
7. Averages
8. True
9. False
10. True

CHAPTER 6: PATTERN ANALYSIS

1. No
2. No
3. No
4. No
5. Yes

CHAPTER 7: PROCESS CAPABILITY

Practice problem 1

a. Frequency distribution for the data

b. $\bar{X}$ = .7518 S = .0027

c. C_p = 1.23 C_{pk} = 1.01

d. Z upper = 3.04 Z lower = 4.37

Percent yield = 99.88 percent

See Figures H.1, H.2, and H.3.

Practice problem 2

a. $\bar{X}$ = 3.35

b. Sample sigma (s) = .0196

c. C_p = 1.28

d. C_{pk} = 0.51

e. Percent yield = 93.7 percent

See Figures H.4, H.5, and H.6.

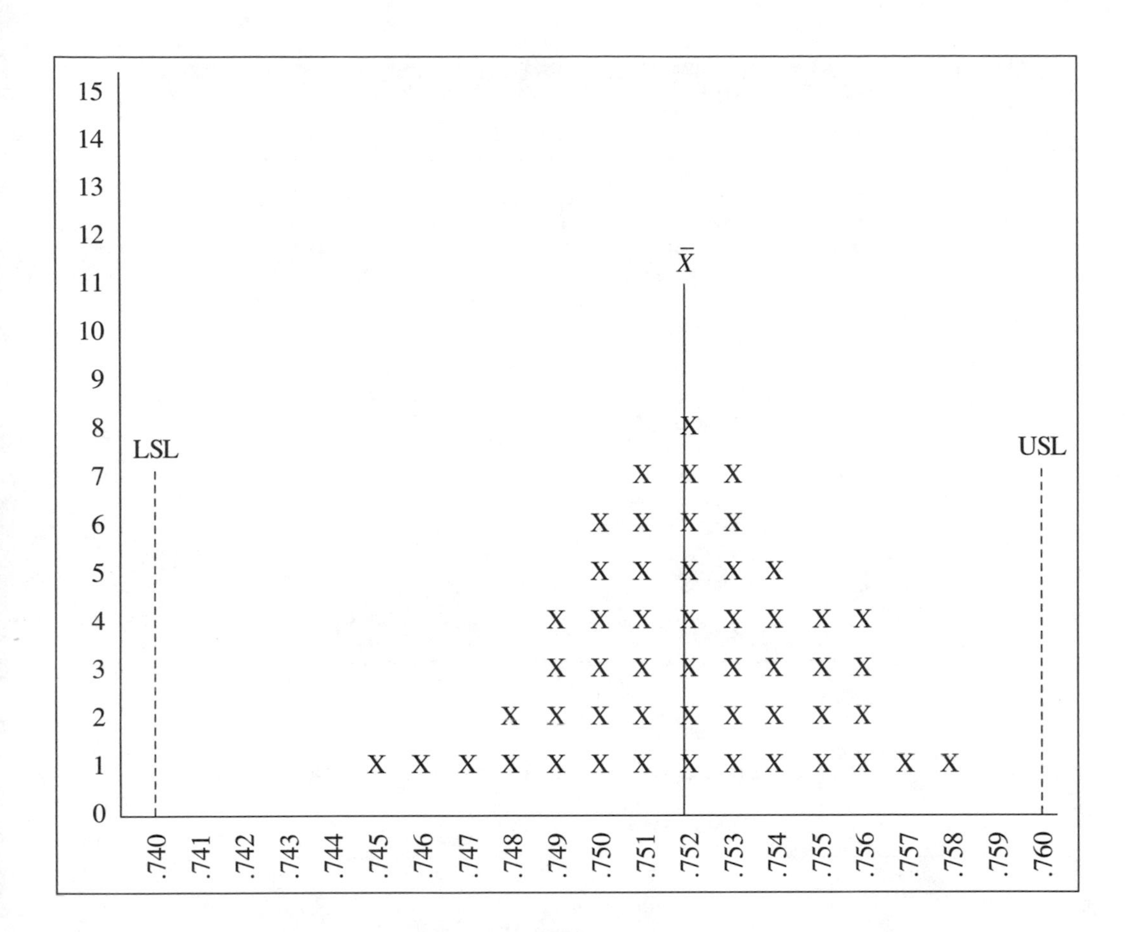

Figure H.1. Solution to capability practice problem 1.

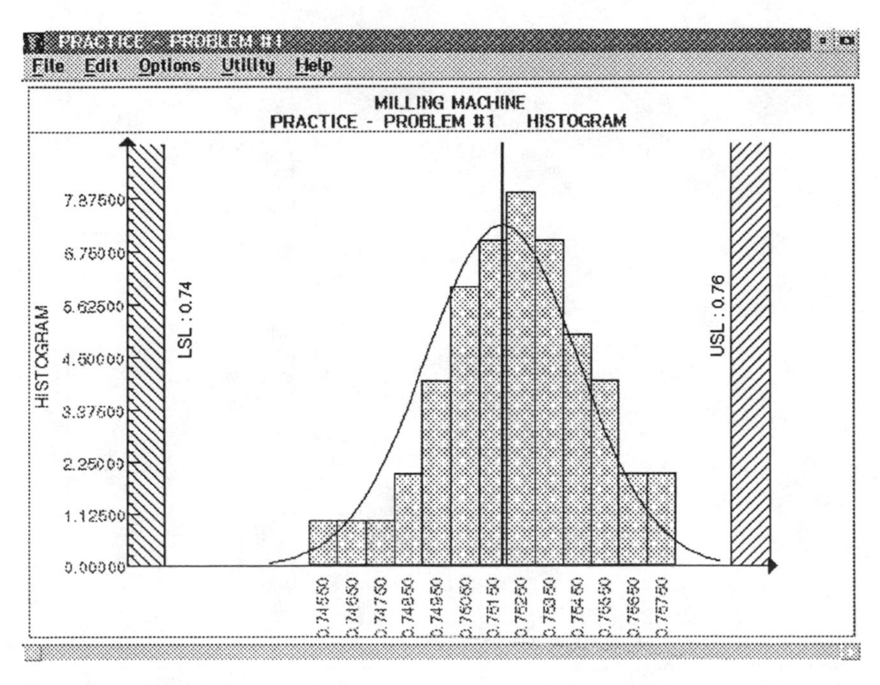

Courtesy of CIM Vision International

Figure H.2. SQM software example of solution to capability practice problem 1.

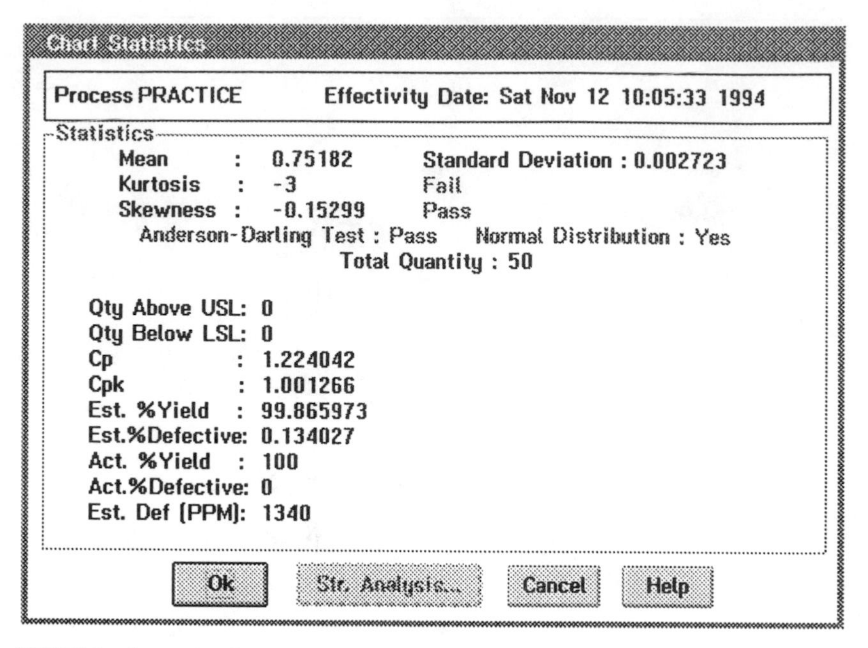

Courtesy of CIM Vision International

Figure H.3. SQM software example of capability report for practice problem 1.

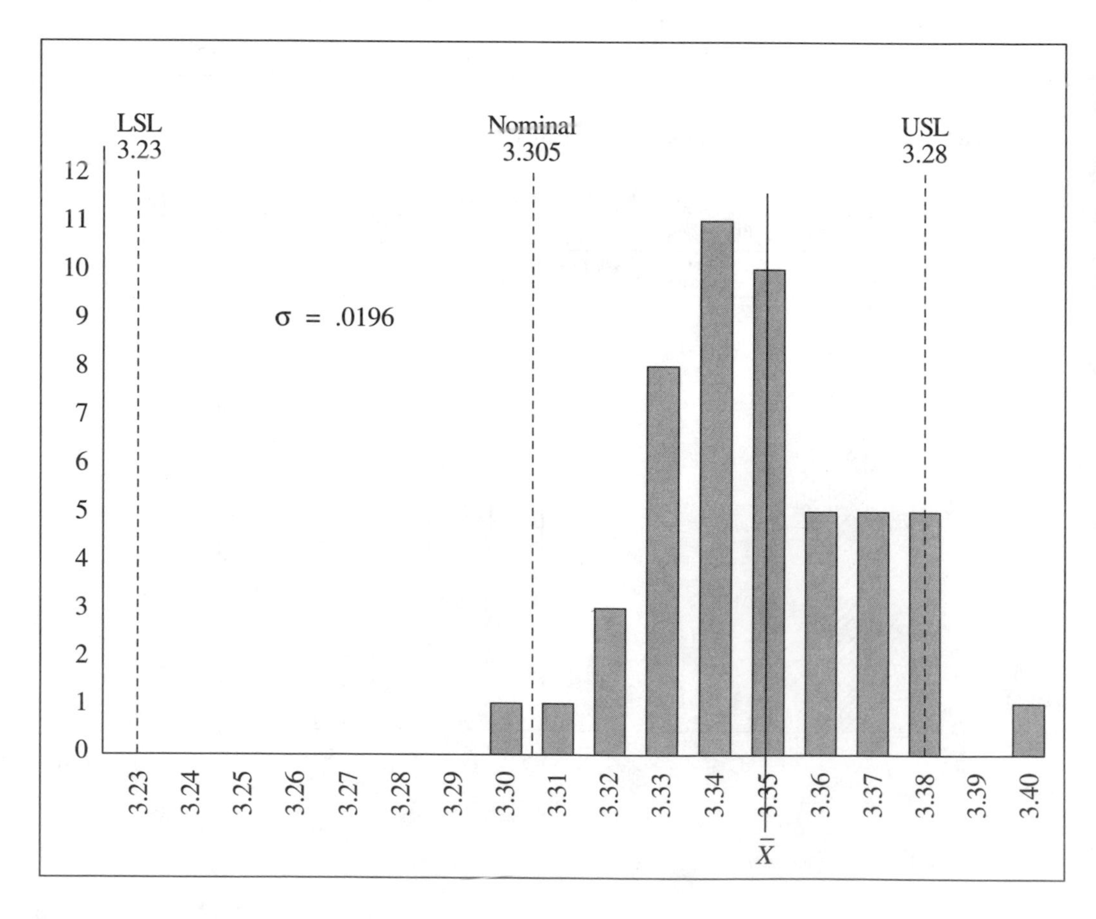

Figure H.4. Solution to capability practice problem 2.

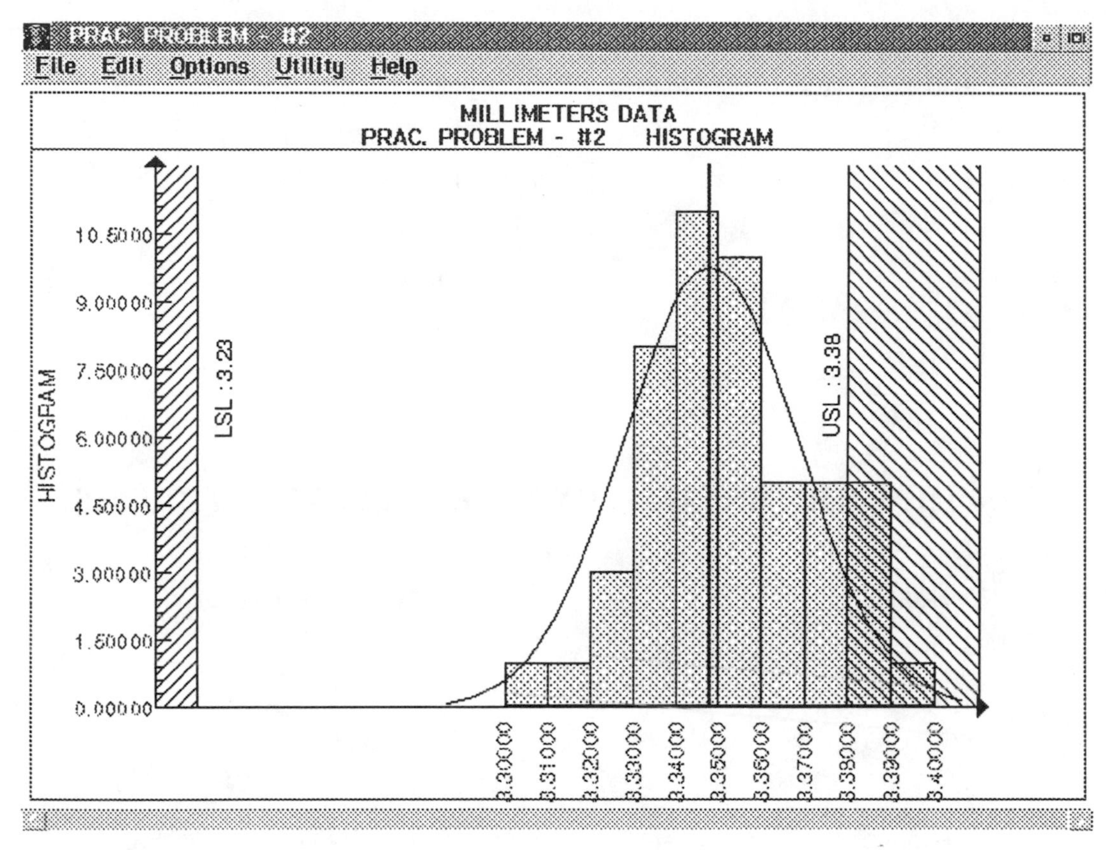

Courtesy of CIM Vision International

Figure H.5. SQM software example of solution to capability practice problem 2.

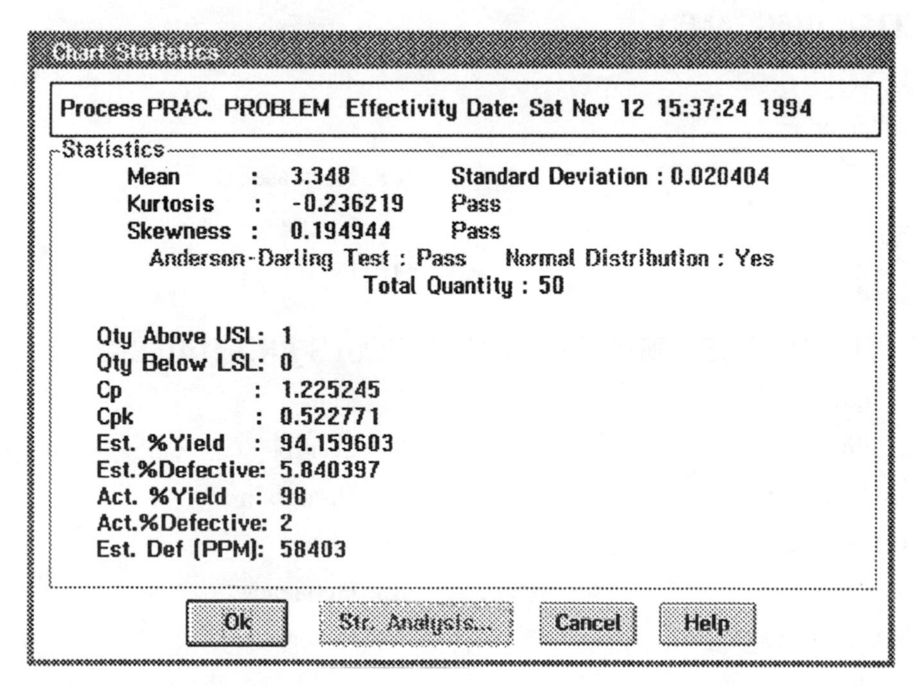

Courtesy of CIM Vision International

Figure H.6. SQM software example of capability report for practice problem 2.

Review Problems

1. 0.93
2. No
3. 0.31
4. No
5. d
6. No
7. 49.04 percent
8. 0.67
9. d
10. 85 percent good parts
11. 2.67
12. Yes
13. No
14. 6.67
15. Center the process

CHAPTER 8: GAGE REPEATABILITY AND REPRODUCIBILITY STUDIES

1. e
2. +.0002″
3. True
4. c
5. Precision
6. c
7. Accuracy
8. Long range method
9. Long range method
10. Long range method
11. No
12. c
13. False alarm
14. True
15. e

CHAPTER 9: REGRESSION AND CORRELATION ANALYSIS (SCATTER DIAGRAMS)

1. X
2. Increases
3. Paired
4. Intercept
5. Coefficient
6. True
7. c
8. Decrease
9. Circle
10. Does not

CHAPTER 10: PROBLEM-SOLVING TECHNIQUES AND TOOLS

1. Root
2. Environment
3. Cause, effect
4. Follow up
5. True
6. Check sheet
7. Flow
8. Matrix
9. Run
10. Control
11. Histogram
12. Scatter
13. Frequency
14. Histogram
15. d

CHAPTER 11: MISCELLANEOUS TOPICS

1. c
2. Parameters
3. True
4. d
5. e
6. True
7. (a) 1.67, (b) 1.33,
   (c) 99.9994 percent, (d) 60 PPM

Bibliography

Besterfield, Dale H. *Quality Control*. 4th ed. Englewood Cliffs, N.J.: Prentice Hall, 1994.

Grant, Eugene L., and Richard S. Leavenworth. *Statistical Quality Control*. 5th ed. New York: McGraw-Hill, 1980.

Griffith, Gary. *The Quality Technicians' Handbook*. 2d ed. Englewood Cliffs, N.J.: Prentice Hall, 1992.

Hayes, Glenn E., and Harry G. Romig. *Modern Quality Control*. Rev. ed. Encino, Calif.: Glencoe, 1988.

Hradesky, Jack, and Wendall Paulsen. "SPC's Missing Link." *Quality* (March 1987): 55–57.

Juran, J. M. *Quality Control Handbook*. 3rd ed. New York: McGraw-Hill, 1979.

Juran, J. M., and Frank M. Gryna. *Quality Planning and Analysis*. 3d ed. New York: McGraw-Hill, 1993.

Pyzdek, Thomas. "Process Control for Short and Small Runs." *Quality Progress* 26 (April 1993): 51–60.

Task Force of Chrysler Corporation, Ford Motor Company, and General Motors Corporation. *Statistical Process Control Reference Manual*. Troy, Mich.: Automotive Industry Action Group, 1992.

Western Electric Company. *Statistical Quality Control Handbook*. Indianapolis, Ind.: AT&T, 1956.

Index